Human Entrance
to Transhumanism

A Tract on the Black Magic of Materialism

Human Entrance to Transhumanism

Machine Merger and the End of Humanity

James Tunney

James Tunney is the author of this book and owns all copyright © in this work and assert all moral rights thereto 2021.

Cover image and inside art copyright © James Tunney.

ISBN: 9798475734766

To Paul Chek

Also by James Tunney

NON-FICTION

TechBondAge
Slavery of the Human Spirit

Empire of Scientism
The Dispiriting Conspiracy and
Inevitable Tyranny of Scientocracy

COMBINED NON-FICTION & POETRY

The Mystery of the Trapped Light
Mystical Thoughts in the Dark Age of Scientism

The Mystical Accord
Sutras to Suit Our Times, Lines for Spiritual Evolution

FICTION

Blue Lies September

Ireland I Don't Recognise Who She Is

Contents

1st Part - Transhumanism and Deconstruction of the Person...... 9
The Impossibility of Ignoring Transhumanism 10
Anti-Human, De-Personal Technology 29
Scientism and Deconstruction of the Human,
    Individual Person.. 46
Imperial Incentive to Transhumanism:
    All Roads Lead to Rome 60
Psychopathocracy or Machine Misanthropy 68
Prosthesis of Transhumanism to Posthumanist Prison 76
2nd Part - The Magic Spells of Transhumanism 100
Reducing Persons to One of Many Conscious Agents 101
Spirit-Mind-Body and Boundaries 117
Technique and Magic 134
God and Golem, Black Magic 150
The Transhumanism Trance 159
The Transhumanism Trick 178
3rd Part - Sensing the Machine-Human Merger 201
Goldilocks, ER and Lagom 202
Kafkaesque 211
Androids Dreaming of Electric Sheep in a Crack
    in the Universe... 219
Eichmann's Children 224
Humanity Rest in Pieces:The Fourth Industrial Revolution 231
4th Part - Re-asserting the Real Qualities of People 239
Quant T.R.A.P to Quantum Q.U.A.L.I.A 240
Servitude, Deconstruction, Incarnation 254
Ante-Humanist Metamorphosis to Animasphere 261
Resist Dispiriting 288
About the Author 289

"What I would ask you is, to show me why, since each new invention casts a new light along the pathway of discovery, and each new combination or structure brings into play more conditions than its inventor foresaw, there should not at length be a machine of such high mechanical and chemical powers that it would find and assimilate the material to supply its own waste, and then by a further evolution of internal molecular movements reproduce itself by some process of fission or budding. This last stage having been reached, either by man's contrivance or as an unforeseen result, one sees that the process of natural selection must drive men altogether out of the field; for they will long before have begun to sink into the miserable condition of those unhappy characters in fable who, having demons or djinns at their beck, and being obliged to supply them with work, found too much of everything done in too short a time. What demons so potent as molecular movements, none the less tremendously potent for not carrying the futile cargo of a consciousness screeching irrelevantly, like a fowl tied head downmost to the saddle of a swift horseman?"

George Eliot
'Shadows of the Coming Race' in
Impressions of Theophrastus Such (1879)

THE FIRST PART

Transhumanism and Deconstruction of the Person

Technology will eliminate the human with its embrace. We cannot now avoid engagement in 'transhumanism' discourse. It is but a thrust of a broader movement of technology, technics and technique. Growth of governance by the converging technological domain, 'technium' or 'technosphere' causes constrictions of human activity, as well as creating apparent freedoms. The human, individual person idea is being deconstructed, largely by scientism or inappropriate applications of scientific thinking and authority. Transhumanism will alter humanity through the notion of mechanical or material enhancement by prosthesis, extension of organs, hi-tech addition to, or replacement in, the human body or mind and cyborgs. Transhumanism may involve a transition ultimately to 'posthumanism.' Such movements represent actual assault on the human, individual person. Proponents seek licence to fundamentally alter us and proclaim it as liberty. The idea that transhumanism is exploratory hides a danger that it is a tool for mass enslavement and termination of *homo sapiens* as we know it. The human body will be isolated, enclosed, invaded, conquered and colonised. The technosphere will become a pseudo-intelligent entity, to integrate, consume, reset, fabricate and make us up.

The Impossibility of Ignoring Transhumanism

"Never underestimate the power of stupid people in large groups."
George Carlin

The Age of Reason has given rise to the Age of Treason. Humans live in a trance, enthralled by technology. Hi-tech science is becoming the religion of religions and seeks to monopolise wisdom. Scientism, which is an application of scientific methods beyond appropriate dominions, has become an ideology. It represents an apotheosis of the Enlightenment, machine-view of the material world. Such a physicalist perspective reduces life and experience to a narrow range of lenses and mere measurement devices that concentrate on quantities and metrics whilst ignoring qualities, metaphysics and consciousness itself. Success of the scientific method manifests in a proliferation of technological systems and the promotion of data-based techniques. Scientism combines with real dominant forces such as cybernetics and computation to focus on control akin to an essentially magical power. Phenomena such as convergence have concentrated technological power in networked systems of governance beyond citizen or state control. Globalism promotes a new 'state of the world' or a 'new world order' through a 'new industrial age.' Transhumanism is one of the purest, purist policies of this distilled materialism. The consequence will be demise of humanity as we know it. Failing to have a sustained view of ourselves is allowing a simulacrum grow to replace us.

We say we want eggs but decide to exterminate hens. We are whole beings greater than the sum of our parts. Scientific method is always the prime, great deconstructor. Philosophical theories of deconstruction are often pseudo-scientific offspring thereof that fail to recognise their parents. Consciousness that produces science has warped into scientism as well as philosophies that are instruments of a scientific agenda perhaps unbeknownst to proponents. Science has achieved great power through deconstruction, breaking, cutting things, dissecting, chopping, opening, going inside, finding constituent elements, digging, extracting, using living subjects, experimenting, testing, controlling, 'torturing' or vexing nature. In doing so, materialism has regarded important resources and results as 'externalities' and mere by-products. Scientists and technologists in their single-minded pursuit of particular objectives ignore other consequences often because of the possibilities of ready strategic, military or commercial exploitation. Fragmented thinking creates shards that damage other parts. Oblivious to responsibility for damage and blinded by power resulting from success of its tools, science and technology are proposed as solutions to unacknowledged problems they create. All the while the 'technosphere' or 'technium' grows. Such terms refer to the entirety of hi-tech networks and systems created by humankind. The technosphere is nearly becoming capable of acting as an organism. We will be subjects of it or IT and assimilated or destroyed thereby or subsist in a wasteland that was formerly a biosphere. Much elevation of machines and technology is associated with a failure to regard the significance of the person as a whole organism with meaning and purpose determined by consciousness.

A focus on 'ratiocination' supported by quantification and machines as the exclusive objective of a 'scientific' society threatens humanity. The idea that mechanical computation is the supreme value and should replace the vital force of human persons is becoming dominant and pervasive. We are entering scientocracy or an Empire of Scientism. The supposedly glorious and clearly superior achievements of science and technology and their obvious benefits are constantly advanced without adumbration of the crippling costs. A blind spot exists with regard to the horrendous mistakes made in the name of science. If some phenomenon is good, science claims parentage; if bad, it is an orphan, or better still can be blamed on religion. For example, study of the daftness of phrenology, craniology and measurement associated with the brain and dangerous racism promoted as authoritative by science, is ignored as merely an aberration. Such science was used to destroy, deny and diminish people with deleterious consequences. Now scientists, along with some of their sycophants, have increased their bet on the quantitative approach. They have become convinced again about their omniscience and ignore any prescience by those who consider more fully. It is not merely some unfortunate race of people fit into an unscientific analysis by apparently scientific schemes that need worry. Science is at war with the human race itself. Increasingly ruled by graphs, models, unquestionable authority and the religion of science, we are sequestered, pending our transmutation by technology. Scientocracy provides an apparatus to destroy humanity. We have crossed into obsessive rule by experiment, data and control with glorification of information and contempt for the interiority, inwardness and integrity of human persons.

Bewildered, you may find yourself in a wilderness. We are giving the reality and tradition of humanity away. We give it away very cheaply, daily. Like indigenous people supposedly fascinated with a bauble into yielding up that which they had no comprehension of being able to, we give up our very humanity. 'Tradition' comes from a Latin base associated with handing over and is apparently etymologically related to the word treason. That we surrender our species through words and concepts as we succumb to machines and technique must be magic of a sinister and sorcerous sort. How else can we explain that we have suddenly accepted our inevitable demise and sacrificed concepts of liberty so readily? We seem spellbound by shallow philosophy. We worship wizards who deploy idols of technology. We are bewildered by image. This wildering will bring us astray. We go down into an abyss of our volition. A winding stair comes from a winding stare when we watch. Giving our attention to the wrong things we lose all. What do we lose? We are losing our humanity and very personhood. Mesmerised we relinquish claims to our personal sovereignty and that of the human itself, and all our ancestors in us and all possible descendants resident therein or societal inheritors.

An assumption of the majority of people uninterested in transhumanism is that they can passively ignore the phenomenon and leave other people play or get on with it. Unfortunately, the nature of networks, 'network effects' and tendency to pervasive surveillance combined with dependency on mobile devices, when linked to the desire to achieve global governance, will create momentum to compel transhumanism. Like sale of illegal drugs, the wicked witch in Hansel and Gretel or Child Catcher in the film *Chitty Chitty Bang Bang* (1968) there is a phase of sweet deception leading to confinement. If, for example, we are inundated with evidence which suggests mobile phones have deleterious medical effects and when people are dependent thereon, fear would make it much easier to force dependents to accept some new 'healthier' method of being networked. If the new method was a wearable technology or especially an implant, then the Rubicon would be crossed. Control of the technical nature of networks with inevitably increased convergence, concentration and huge network-dividends translates into commercial power that transmutes into political potency and ultimately even into governance. Command of market power beyond reach of any effective regulation as a result of institutional capture and transnational flows of capital, creates possibilities for imposing transhumanism. Such compelled transhumanism need only be a small step above prior levels of network dependency. If, for example, the force of Big Pharma can continue to impose its will on global governance associated with public health and if it can combine with Big Tech to ensure some bio-metric or bio-tech admin system that can be projected on the world, then the first and final steps to global transhumanism will be achieved. As smart dust and nanotechnology increase, temptation and tendency to use it for supposed security, health and other ostensibly beneficial reasons, becomes irresistible. Restraints on the evolution of such systems are weak. The propaganda model really only means that she or he who pays the piper plays the tune. Big Pharma control much media through advertising. Big Pharma and Big Tech operating beyond the power of the nation state and in the absence of international laws, can effectively project a system of global governance based on hi-tech admin of bio-security. This is the basis of globetechgov. Big Pharma and Big Tech are crucial, global forces in the construction of the new world order. Bio-Tech global governance guarantees that transhumanism does not merely remain a marginal or sub-cultural, geeky activity. Transhumanism becomes a central control possibility in the new architecture of global governance based on Big Bio-Tech. Propelled by any convenient pseudo-public health justification, promising massive return for those who promote the agenda, transhumanism can be sold as sensible policy through the same propaganda machine controlled by Bio-Tech. We will have no choice soon in a context where our currency, money, communications, social benefits, health services and education are effectively under power of corporations beyond control. They are the puppet-masters and all other institutions are puppets. We are minor, external dummies controlled by the system and our behaviour increasingly programmed by a policy perfected over decades by the best behaviourists, cyberneticians, psychologists and psychiatrists that money could, and did, buy. We become mere data objects.

I remember my grandfather using a hand-held plough. I remember seeing tv for the first time. I remember seeing mobile phones for the first time. In a generation or two, change can easily become revolutionary. Change can be revolutionary because we are experiencing a permanent, cultural revolution. This revolution is not the result of organic technological development. It is rather the logical consequence of an ideology that is creating the Empire of Scientism supported by an information cult. Not only is this a tyranny but it is a totalitarian regime. Not only is it a totalitarian regime but it is a total totalitarianism. Never before could such a system have worked. It was not that Hitler, Stalin or Mao would not do it, but they never had the means. Now we face followers of all those types unifying to manage a tool of totalitarianism. The blueprint of the thought-process was suggested by H.G. Wells (1866-1946). His books proposing a 'New World Order,' which would be wrought by an 'Open Conspiracy,' were based on destroying every other loyalty until there was nothing to prevent the rise to power of the scientific elite, their objectives and supposed reason. It is clear that this wizardry is of a Nietzschean type that Sri Aurobindo (1872-1950) anticipated. The demolition of restraints is not to get to reason but simply to give utter or 'absolute freedom.' This is often what psychopaths seek. J.D. Bernal (1901-1971) demonstrated that there would be no moral restraints. Curiosity, control and the docile people would be subjects for experiments by the scientific elites.

Transhumanism is sold by the thin edge of the wedge. You will be emotionally blackmailed by legitimate and joyful medical advantages. You will be frightened by the inflation of some risk. You will be cajoled and coaxed by non-entity celebrities. All of the propaganda machine may be orchestrated so that you must adopt whatever new implant or nano-bot they come up with on some concocted, cock-and-bull basis. Like as with the 'war on terror' media can wield its enormous power to woeful purposes. As with massive deception used to force an unwilling public to accept the supposed necessity for infliction of massive pain on peoples whom they had no quarrel with, when a megamachine wants a result, it will get it. Most policy covets power and proceeds covertly.

You are the colony now, the object of colonial power. You are the resource to be invaded or enslaved. Massive, technical experience in exploitation has been distilled into the most potent form to fuel frameworks for the greatest enslavement ever known. Technological networks are key. Transhumanism is an obvious tool in the totalitarian box. The tech-Hydra extends tentacles towards our minds touching everything around as a technosphere traps. Neutral positions on such ideas as transhumanism, the 'Singularity,' conscious agency or posthumanism, may represent acceptance by acquiescence. The clean, jolly, optimistic, progressive, libertarian, helpful front of transhumanism conceals a disturbed back. Scratch the surface and you can see deep, anticipatory defensiveness that threatens those who oppose it. The argument that transhumanism is only about self-expression is dispelled by arguments about issues such as dismantlement of kinship. Other ideological strategies indicate it is never about mere technological enhancement. It is sometimes an ideology or part of the ideology of scientism. As such it is a total solution and source of opportunistic revolution. It cannot be ignored as something to be left to geeks and sci-fi aficionados. The soft-focus shiny sale of transhumanism has a sharper side. If you do not agree with extending or enhancing humans by deconstructing them or if you do not recognise non-human persons, you are a 'human racist.' Techno-futurists and techno-utopians want to decide what is best for you. It is not the individual 'body' or 'bio-hackers' or 'scrapheap transhumanists' that present the most problem but concentrations of power that seek to commercialise or otherwise compel the masses. Artists perceived it, scientists planned it. Philosophers warned us.

If we merely look at how the motor car took over the world in a century we see how technology impacts. It has benefits and costs. Costs are environmental, cultural, social and personal. The car-infrastructure is a machine that makes malls and can turn paradise into a parking-lot. The fuel that is life-blood of this particular and general machine has caused wars. We imagine we are masters while we must serve the network necessary to make such machines work. As robots proliferate, we will find ourselves serving the system that makes them, until we end up serving machines. As technology around us engages and entraps us with wiry ivy, tendrils and lianas, we will be transformed into nodes knotted into networks.

Furthermore, since humankind first picked up a stick to use as a tool, we have created extensions of our body. A tool or prosthesis extends organs as thinkers like Ernst Kapp (1808-1896) explained in *Elements of a Philosophy of Technology* in 1877. As Freud (1856-1939) and others indicate, prosthesis is an important explanation of human nature. Since we concentrate so much on tools and adding things to lives and bodies, we then create conditions to be destructive to the race itself. This leads to a stark choice between nature and culture, the organic and inorganic, between machinery or lively vitalism, impersonalism or personalism and mere behaviourist explanations of people or philosophical and metaphysical ones. We face focus on attributes or individuals. We face a choice between parts or wholes. That is why some work of thinkers like Franz Brentano (1838-1917), Kurt Goldstein (1878-1965), Hans Driesch (1867-1941), George Canguilhem (1904-1995) or Edith Stein (1891-1942) is relevant and the disposition and discussion of personalism and holism is meaningful.

Such thinkers look at humans as complex beings that are not reducible to attributes by norms and pathologies projected on them. Mechanistic explanations do not reflect the full picture. The human, individual person is a whole organism comprising the body, mind and independent, spiritual consciousness. They are special and unique with access to the power of imagination. The future should be fashioned respecting individuals and protecting against persistent predation by groups who covet power. While we must address our relationship with animals, plants, earth, water, air and other people, it cannot come through compulsion and enslavement. It is necessary to see clearly what technology actually is. As Kapp and others indicate, we invent first and only thereafter recognise unconscious motivations which made it happen. Like technology reveals itself, it reveals us. As we grope around in a world of stakeholder-algorithmic-propaganda-marketing-driven relative 'truth' anything can be made sound persuasive in a repetitive familiar, jingling or juggling way.

There are maybe a dozen strands in personalism and some, such as African or Latin American, have not figured as much as they should. They all tend to reflect a core of ideas. It may come about because of persistent belief in a manifestation of a higher order or consciousness in us. The central idea is about significance of the person as subject and as an end that is more than material and not dependent for being on the structure of society. Associated with this is an important idea of the respect for others through relationship in 'I-Thou' or face-to-face contact. This can be seen in the work of Martin Buber (1878-1965), Jacques Maritain (1882-1973), Emmanuel Levinas (1905-1995) and Denis de Rougemont (1906-1985).

Most personalists feared repression of personhood itself through totalitarianism. Machine-thinking and de-personalisation are critical approaches in totalising control systems. Many personalists saw that technology attacked relationships between different persons. They often perceived the connection with control systems and rights. The origin of personalism for some is linked to Friedrich Heinrich Jacobi (1743-1819), Friedrich Schelling (1775-1854) and also Hermann Lotze (1817-1881). Some personalism sees significance in revelation and says that reason, rationality and utilitarianism emerge from other conditions and that values must be considered as well as mechanism. Anyway, you know you are a person that *is*. Everything you know and can know comes from that being and knowing that you *are*. But others may make you a means to an end with disintegrational doctrines and thus disregard you through the lenses of attachment to some principle of power that reduces and ignores you.

We cannot ignore transhumanism as a phenomenon predominantly inspired by science. Cybernetics influenced thinking on human behaviour and nature itself. Success of such techniques through ratiocination is displacing other analysis. Biology is enlisted and neuroscience reinforces this reduction. An exclusive belief in quantification and calculation, as part of a science religion with an ideology of scientism, is creating a reduced idea of human value. An inflated view of calculation's significance promotes the superiority of machines, technology and technique, to reduce humanity by mechanisms. Biology and nature are seen to be merely failed mechanisms needing to be further enhanced to reach predictability of machines. Scientific technique is infused into corporate personality.

The technique to transform humans into manageable machines involves a cheery carrot of enhancement and a sharp stick of compulsory compliance. Unlike libertarian or anarchic morphological 'freedom' suggested sometimes by very wealthy people, we plebeians will most probably experience such forced, mass, technical intrusion and interference in the body as is necessary to colonise us. We will be easily colonised by computational power through mass acquiescence to technology, making us dependent for all necessaries. Boundary breach of our sovereignty by technology creates bridgeheads to end our sovereign consciousness. We cannot ignore the possible anti-human, deconstructing, de-humanising, de-personalising potential. We must be clear about what we conceive ourselves to be and defend who we are, as others seek to disestablish us. Long-established rights protecting us against arbitrary power are being eroded. Ontological inability to reduce alive, vital organisms and people with mere mechanistic explanations alas does not prevent viral techniques and technology from ousting living systems. You will be increasingly regarded as a messy machine with destructive impacts improvable by technique and tech-addition to replace such elements unlike a machine by reducing flesh, blood and free-will to a fiction. Transfixed, we are being daily and alchemically reduced from human subjects into enslaved, technical, networked objects. This is no mere hypothetical, speculative, paranoid or science fiction exploration. The reality of the mechanisation of society, nature and humans has been a policy pursued with vigour and determination. Those who might be appalled by the true implications have been charmed with cheap charisma. The present crisis is feeding frenzy for mandarin piranha.

Difficulties making assumptions about motivations of transhumanism come from similarities between the aims in quests for spiritual evolution and techno-transcendence of physical limitations. This is most clear perhaps in the trajectory of development of types of transhumanism in Russia. There is a strand which blends Russian Orthodox Christian philosophical speculation and technological development with some left wing, utopian ideals. This is still relevant in contemporary Russia. In particular, the idea of resurrection becomes a goal in this dimension and communication with ancestors and the realms of the dead becomes transformed into a technological project. Thus parapsychology could fit into such a techno-worldview. Similarly there seems to be a stream in Mormonism which accommodates transhumanism. Mormonism also has a retro-encompassing of ancestors that had been important in the thinking of certain Russian philosophers. It is noteworthy that the founder of Mormonism, Joseph Smith (1805-1844), had an interest in magic. Scientology was founded by an individual deeply interested in magic. These interests are not as rare as people think. Our habit of putting things into neat compartments is part of a deeper belief in the power of scientific taxonomy. It is important to be open-minded in order to comprehend the powerful forces that shape transhumanism. This involves some consideration of science, religion, theology, magic, politics, law, philosophy and psychology, as well as specific study of phenomena of *inter alia*, technology, commerce and governance. We are still, nevertheless, shepherds of our own dreams and marshals of our own messages as we present evidence. However, we are being enslaved and altered so that the human race will end.

Transhumanism may be part of something else. It is a head of a digital Hydra that reappears in different forms. The underlying force involves a hatred of humanity and a love of machines, metal and material. A desire for power drives this. It encourages violence, mechanisation and an assault on nature. The essence of this monster derives from the concentrating force of instruments, push-button-power and keyboard control. Nature and humanity is to be literally sacrificed to the lust for power that technology gives. Violence, speed, youth and war are cherished as goals that see humanity vanish. This is not my fantasy but the thoughts of the Futurists that inspired both Fascism and Communism, in the work of people like Marinetti (1876-1944) who published his *Manifesto of Futurism* in 1909. They are described as nationalistic but really were imperialists. They were serving the emergent Empire of Scientism. Fleeting sensation from machines replaces your feeling, companies replace compassion, huge machines will replace humankind. Do not project onto the drivers of society values that are not there. Do not presume restraint of the type you may assume. Do not expect reasonable limitations to what can be done to you or your family or friends. Reason is merely an instrument. Your present is becoming the future that certain artists and political players projected. Their magnetic attraction to machinery has magnified by multiplication through our involvement. Economists, lawyers, politicians, psychologists and many philosophers have generally failed us. Many of these disciplines have been emasculated and turned into servants of totalising technique or ideologies promoting them. All they need sell you to succeed is a sweet fragment of the poisoned dream.

You will be told that humans have always been technology or machines. You will be told that you are to be upgraded like machines. That is what has been happening to you but you did not know it. You will be told that all is normal. All the efforts to insist that you change physiologically are reasonable. You were a cyborg all along and you did not realise it. You are already technologically-altered organisms they say. You will be told that everything is psycho-physiology. You will be told that the immaterial element that you thought existed was an illusion. You have technology such as language in you. You must see gene-editing as merely like changing language. You will be told that this is just non-dualism. They take cybernetics and use it to re-write human history. You have no autonomy. You must now adapt to new technology and new thinking. The studies and scholars often make arguments that become policy and this supports scientific corporations. You cannot ignore it and should not. We are in trance, entranced to transhumanism. You do not have to agree with the personalists or spiritual people about the origin of the person, nature or existence of the divine as the basis for protection of the person. You do not have to look to religion as a sole source. We should rather look at the presumptions we may have made and foundations we assume about the person and realise that they are crumbling. As people let institutions wither they may also let some valuable perennials go with them. You will not miss them until they are gone. If we let that happen we must take the consequences. Just in case it is not obvious what is happening, let me spell out the spell cast or broadcast upon us.

The promotion of transhumanism will be projected on the basis of helping people. It will be sold on the basis of the freedom of the individual, liberty and the economy. It will claim to represent all science and technology and try to suggest that every medical and scientific advance is somehow related to it. It may label opponents as violent, backward thugs responsible for all the badness, mistakes and mischief in the world. The bio-conservatives will be mean-spirited people who really should not be entitled to use technology or expect medical help because all such benefits will be somehow artificially attributable to transhumanism through their re-writing of history. The narrative of enhancement will be elastically applied to examination of evidence. The evolutionary 'rights' they advance to add to cryonics, cloning, genetic engineering, bionics and stem cell therapy will correspond with their attack on the rights of others. The individual who ignores the claims will soon find out that the agenda is one of compulsion and the object is to entrap them in the network through implants until all are transformed save perhaps the transformers. The nature of global governance is that it is converging around science, technology and control and aims to be totalitarian. This agenda is a scientific one and the movements manifesting on this trajectory are predictable. Humans will lose out because people who love 'liberty' and machines and their own existence will sacrifice yours. Their desire to achieve their power and immortality will be the ultimate selfishness and have no attachment to scruples. While chiding your morals they will be hiding theirs. Science will help saying you have no free will, are a selfish gene, a bunch of neurons, blind to 'real' scientific reality, merely an association of attributes.

There is some idea that capitalism and real, free markets in the West are the cause of all problems in the world. Another opposing view is that the model makes sense. Now we are told the system must change, often by the same people who drove the failed market tumbril. Management, technocratic elites and hi-tech companies in the military-industrial, crony capitalist system captured most significant institutions. If institutions cannot be captured, new ones are created. International bodies cut off from direct democratic base act in an integrated way. Abuse of financial and trading systems through legislative capture transfers more public funds to converging power bases. In competition law, the fact of coordinated activity is sufficient evidence in certain cases to give rise to a claim of conspiracy. The degree of co-ordination in centralised commercial activity of corporations using cybernetics and cyberspace across the world and parallel decrease in rights of citizens is clear. We face coalition of these two dysfunctional, materialist systems. Communism and crony capitalism will unite through their shared desire for planning and predictability in a system of global governance with great rewards to the cadre who capture global governance. Success of scientism and technocracy that conditioned this coalition points to possible success of transhumanism. It combines supreme selfishness and the self-death-dread of an arch-physicalist with practical means for the controlling-elites, unrestrained by ethical or moral limitations, allied with technocratic, existential or supposedly evolutionary 'philosophy' that suggests an unlimited capacity to pursue its policy fuelled by an almost mechanophilic, mechanistic misanthropy.

Critics of transhumanism have been working for years now. Fukuyama is one example of a 'bio-conservative.' This term reflects an attempt to project existing left-right terminology onto an evolving context. However it is interesting that progressivism embraces many left aspirations as well as many big business and billionaire neo-liberal views. The main critique of humanism reflects standard left wing critiques from contemporary feminism or Marxism. Many fingers pointing at white men ignore the scientists in colonial powers who invented scientific racism. Curiously however the arguments from people one might expect to criticise the massive concentrations of capital tend to support further applications of this power onto everyone. The anti-globalist protests were re-directed miraculously into pro-globalism sentiment. Even standard arguments about protecting the worker have vanished. The monstrous right seems to be illusory and now is used for anyone who does not agree with the re-directed left. All this seems like a mere game played on the shadows of the cave by our masters to make us ignore the scientific-technical-expert elites who control us and pull the strings, playing political tunes at will with psychopathic abandon. The drive to totalitarianism comes from both left and right, from the spectrum of those interested. Its power comes from technology, technique and science. Its brainwashing comes through the media it controls. That political activists are arguing for Cyborg and Mutant rights when people are losing theirs should make people wake up. Big Pharma supposedly can save us, incurring no liability, with the risk underwritten by the public and message reinforced by Big Tech whilst research rules are abandoned and their power increases inexorably.

So we face a determined group of people with a desire to alter humanity in various degrees and diverse aims. They often state that they want super-intelligence, super-longevity and super-happiness or super well-being. Effectively, transhumanists want to be superhumans and in that sense be above the normal or extraordinary. They want to be very intelligent, live very long and be very happy. I do not oppose genuine private, personal and ethical goals to achieve such purposes. Likewise, I do not think that an equality objection to their ambitions is necessarily a useful or even sustainable one. My main objection is primarily based on the distinct possibility and probability that this movement represents an endeavour that is really calculated to promote mass, networked transhumanism or will be hijacked for that purpose. Implantations that become compulsory for benefit of administration represent not only the greatest threat to human liberty but also pose a danger of extinction of the human race as we know it. If humanity is fundamentally altered, then existing humanity will become something that it is not and be more machine-like. The philosophical-political experience of the search for the super has been a spur for the pure extremist in the last century and a half. Are you ready to allow humans become mere things or nodes in networks? You might ignore transhumanism but it will not ignore you or your family. I use the motif of black magic in relation to materialism, not to vilify transhumanists personally, but because it fits in with a wider conceptualisation of where we are. Transhumanism and posthumanism attack a reduced view of humanism, a straw humanism of a much later type, which ignores the transcendent because of a materialist belief.

Anti-Human, De-Personal Technology

"To be or not to be that is the question."

Shakespeare, *Hamlet* (1603)

As a society, we are in a speeding vehicle of deadly Enlightenment and materialist making. We are now fast approaching a wall or a cliff. Neither seems a desirable outcome save perhaps for the crazy, whooping driver. The short-term power thrill of such a machine makes drivers oblivious to the desirability of restraint. There has been a constant battle between the two mad materialists in the front. Whether left or right, communist or capitalist, individualist or collectivist, they are all still materialists. Sometimes they switch. But the front seat occupants have fought and the vehicle careered through crowds. As passengers we are told that the only solution is to change our form into a dummy to deal with impact as acceleration continues. Breaking down is seen to provide great promise for a future vehicle. You ask whether we could just stop, or slow down or get out to look at the flowers and breathe the fresh air. Then we remember that the drivers love *Crash* (2004) and that it was something about a fetish for car 'accidents.' A thought drifts in as the growing wall looms before the loons.

> *"We're always behind this metal and glass. I think we miss that touch so much, that we crash into each other, just so we can feel something."*

People driving may drift off and drive on auto-pilot. Our society is on cruise-control in the wrong direction. Jacques Ellul (1912-1994) and his good friend Bernard Charbonneau (1910-1996) were among some of the most perceptive critics of our use of technology or perhaps its use of us. They shared the concept of significance of the person. Robert K. Merton (1910-2003), in an introduction to *The Technological Society* (1964) Vintage Edition, explained some of the concepts in Jacques Ellul's book. Merton repeated that Ellul's concept of *technique* was wider than technology and included *'any complex of standardized means for attaining a predetermined result.'* The 'Technical Man' focuses on objectives, concentration on results and deliberate and rationalised conduct. Means improve but ends are 'carelessly examined.' Centralisation and concentration constantly increase. Technique is the totality of techniques and technology. Merton argued that this was some latent or inherent result from the nature of technology and technique. Arguments of technological inevitability can become a type of hi-tech determinism or fatalism suggesting that thought-process and technology create their own momentum. Whilst true, it is a mistake to ignore the vortex which pulls a range of individuals to power-possibilities that result from increased force and tight focus of technology. Machinery power or technique increases in a machine of governance so social and human inputs decrease. The nth degree of a machine mind-set is the incredible policy idea now becoming a quite high probability, that the human merges or becomes a non-biological machine and thus a better version than what they are conceived of as mere machine-objects. A de-personal, de-humanising force follows technique.

To entrance is the entrance. We enter in trance. We go down into Dreamland and Ethereum. We are in a trance, enchanted, entranced, hypnotised, spellbound. In the long experience of humankind we have never been subject to such a continual assault on our perception. Yet we have come to see it as normal. We have been conditioned to accept bewilderment as routine. A monstrous mammon of machinery manifests a maya or matrix which has confused and disorientated us. Screens scream for our attention. We are becoming poppets, puppets, effigies, fragments and figments re-created in the mind of network controllers. We have learned to dance on command as our strings are pulled as lines from puppet-masters determine what signs of submission we must show. Online, in a web or net our lives increasingly determined by links and chains, we are willing to release all our personal sovereignty having been disenchanted and subjugated so successfully that we are ready to become assimilated into the network itself. The illusion of individuality has been promoted as a device to enable us to abandon it. Now prepare yourself for the end of the road of a long heritage that takes you back to the first man and woman. In our hypnotised state, we are primed to allow final transference of our surviving beauty to a machinery of reduced rationality. If we refuse to involve ourselves actively now we will rapidly find that our acquiescence represents our consent. We are accosted by technology every day, more and more. It is a new revolutionary state of normality whose insidious systemic insistence is instantiated every instant. Being accosted we come to it or it comes to us. We must face it literally. The option to turn away is dwindling. The option to opt out of the network of technology is gone in many ways.

Certainly there are many advantages and benefits that can be cited. We hear of those happy forces all the time. But significant, severe costs are seldom promoted. A comprehensive system that manages itself in a largely autonomous way as a new system of global governance can impinge and interfere in a cavalier fashion with rights or expectations of those who choose, or are compelled, to use it. Proponents of such a system cannot be expected to promote the costs thereof. This is especially so if the controllers are totally convinced of their own benevolence or worse have total disregard for those who submit to them. Technology has costs. Technology has costs in terms of the human, social, environmental and economic contexts. Social capital is undermined by technology through interference with local relations in favour of distant and virtual ones. Many technologies are damaging to the environment considering the costs of networks, materials and consumption resulting therefrom. Other techniques such as ingenious farming methods or low-impact systems are depleted when such opportunities are foregone because of a dominant system.

The human species is under threat of extinction primarily because of its infatuation with and worship of the machine, technology and technique. It has fallen in sickly love with a shimmering tech-reflection. Having fallen in love with a pale imitation of itself it has yielded its future to it. Humanity has allowed its fate to become determined by technology and is disintegrating in its will to survive. It is surrendering without a whimper to a story which has become its destiny. Humanity must become something non-biological and messy biology must be replaced. We may first be integrated into the technosphere or 'technium' and then totally assimilated thereby. This domain is the entirety that may become an entity that increasingly controls our lives and compromises our humanity. That which was meant to be liberating has imprisoned us. In particular, transhumanism proclaims the end of humanity. *Transhumanism refers to the disposition, desire, opportunism, philosophy, practice, politics or ideology to enhance the human body or its mechanisms and thereby reduce or abolish its perceived physical limitations, defects or disabilities, re-define personhood or in other ways transform or increase the human through the application, engineering and integration of technology or technique and quantification to human biology to ultimately become a patterned identity, machine or transcend physical boundaries, our species or its existing nature.* Mere restoration of optimum function through technology is distinguishable from enhancement beyond normal expectations. We are at a fork. For some it is *homo sapiens* or a new spiritually evolved *homo spiritualis* on one path with a greater re-integration into nature and a sense of love. Alternatively it is AI, sapiens, cyborgs, androids, post-humans or conscious agents on the other. Unfortunately, one path is likely to destroy us. The Transhumanist Technological Reformation will split society as soon as implications become evident.

Transhumanists essentially want to limit limitations. They want to remove boundaries and restraints and supposedly preserve some element of essence in more persistent, mechanical forms. The will to do so suggests a desire for certainty. Rational predictability and technical improvement become technique which then leads to crystallisation of the will to change the corporeal element of the human as well. An extreme need for certainty is crystallised into a system that seeks to limit uncertainty to only such choices as can be made by the controllers thereof. For the vanguard of transhumanism, the argument apparently revolves around 'morphological freedom' and the desire to de-construct limitations to development. Someone who wants to persist in the standard state of humanity is now a 'bio-conservative' and may be lumped into a category which makes them appear backward, anti-scientific and unreasonable. The wider idea of decreasing, dissolving or deconstructing boundaries means that the argument about technological enhancement becomes an ideological battering-ram that breaks down much more than is required for mere facilitation of tech-enhanced humans. The transhumanist agenda may be able to unite the extreme materialism of the left with the materialist, libertarian right just as many of the neo-liberals had Trotskyite links. The term 'progressive neo-liberalism' has been used to indicate the crossover. This might be called the materialist ideological switch or pendulum. It is clear that opponents of the technological society also realised that it was not an issue of political polarity.

If the path to artificial evolution continues, then the world becomes a total technosphere where the biosphere in any meaningful sense disappears. The 'gray goo' possibilities become real as does the 'Solid State Entity' or the idea that the world is run by an Entity that is made of machines or AI. The old humans like us cannot be tolerated in this scheme. The other path represents the only viable alternative. It is not more of the same but represents a focus on spiritual evolution of humans. It does not preclude technological enhancement but would focus on the physically weak and disabled to restore to a base line where practicable. Spiritual evolution would address destructive tendencies and over-reliance on the materialist paradigm. It would seek to build on the ancient perennial philosophy in a cosmopolitan fashion. It would be nice to believe that there could be a balanced position in between. There is no reason why a spiritually-evolved race would not use the best technology. The difference is that such people would not want to be slaves to mechanical systems or translated into other substrates.

The dream of technological freedom is an illusion threatening to become more nightmarish by the minute. The mass of humankind has been entranced to believe they 'have never had it so good' to live in such a hi-tech paradise. The past was a bad memory and the future is space travel and stars. In a hypnotised state in the Empire of Scientism the individual in techbondage will yield more and more liberty to some scientific incantations or tech spells. The propaganda model serves the system of mesmerism to achieve objectives. Such technique is what the great esotericist Manly Hall (1901-1990) warned about as black magic. In his great work *The Secret Teachings of All Ages* (1928) he expressed a concern about the scientific disposition and linked it to a potent, destructive tendency in magic.

> *"The most dangerous form of black magic is the scientific perversion of occult power for the gratification of personal desire. Its less complex and more universal form is human selfishness, for selfishness is the fundamental cause of all worldly evil. A man will barter*

> *his eternal soul for temporal power, and down through the ages a mysterious process has been evolved which actually enables him to make this exchange."*

We have been bewildered with weird, wired and wireless strands of manipulation. Our puppet-masters control us with networks of great power over our psychology and physiology. Manipulation is for our destruction. The mythic truth behind the vampire, zombie or some nefarious assault on our body, mind and spirit is now a technical reality. Because the imminent attack on our humanity seems authoritative, reasonable, sensible and scientific, it somehow seems less threatening. Above all, it is measured. The scientists do not need to raise their voice, just like Nurse Ratched in *One Flew Over the Cuckoo's Nest* (1962). They speak in measured tones. But most of all they measure. They quantify, control, assess. This is the essence of power in the Empire of Scientism. It measures all it can and disregards that which it cannot. If it is a 'mere' quality or experience such as love, then it is immeasurable and thus immaterial or can be inverted.

In the 1920's, awareness of the potential of atomic power indicated possibilities of species-suicide. This was a very plausible one in the context of the aftermath of World War 1. Awareness of how armaments, designed and produced by scientists, could cause destruction led to a counter-movement within science. Instead of destroying the world and them, science could assume power and rule us in a scientific way after corporations advanced the agenda of science to a place beyond national control. While many talked in terms of peace, it was clear that the movement would persist without the mass which were docile and incapable of understanding the complexity of science. 'Scientific socialism' and Western science within capitalism in corporate form would achieve power on a global scale and exercise enlightened governance through biological bodies. This was articulated by H.G. Wells and elaborated by Bernal. Others like Julian Huxley (1887-1975) shared some of these values. Arthur C. Clarke (1917-2008) agreed. Out of that milieu, critics like C.S. Lewis (1898-1963) emerged. He perceived the danger of scientocracy. This danger was not hypothetical but a very real and open agenda from people he knew and feared.

The tech-critics have been around for a while and do not get the attention they deserve. Jacques Ellul foresaw the problem in *The Technological Society* (1954). He foresaw the illusion of freedom provided by technology. He said that technology was regarded now as sacred instead of nature itself. His friend Bernard Charbonneau identified too how technology was associated with consumerism and entertainment and a loss of freedom that would lead to totalitarianism. Ellul understood that technique and technology displace other forces and tend to totalitarianism. Both communism and capitalism are committed to the industrial, production-line model associated with Fordism and Taylorism. Heidegger (1889-1976) understood how technology was a force of itself on its own to be revealed. Spengler (1880-1936), Mumford (1895-1990), Jonas (1903-1993), Ghelen (1904-1976). Adorno (1903-1969), Horkheimer (1895-1973), Weber (1864-1920) and Marcuse (1898-1979) also saw dangers. Whatever analysis proves apposite, opposition cannot be predicated on vastly asymmetric and futile violence.

Violence is not a solution and personalism should not permit it. David Skrbina wrote a book based on his conversations with Ted Kaczynski, building on his philosophical awareness of writers like Ellul that was called *Technological Slavery* (2010). This dealt with Kaczynski's Manifesto which was the basis of his violent campaign as the 'Unabomber.' Such violence is not part of the analysis of the French thinkers. Skrbina elaborates on the philosophy of technology. Skrbina seeks to distinguish the critique from the action. Essays like 'Why the Future Doesn't Need Us' in *Wired* (2000) by Bill Joy echo these fears. The argument relates to the lack of connection between our natural evolution and later technological evolution. In a Procrustean attempt we can destroy the earth and make humankind fit into the technosphere. Forces forward increase the gap with our natural state. The momentum leads to destruction through technological determinism. My critique hitherto has focused mostly on the governance issue. Technology is the ideal tool of total-totalitarianism. However, the drive towards technocracy creates other dangers. It could be that tech-determinism inherent in technique means that totalitarianism is an inevitable result. None of these potential dangers are in the interests of humanity and may be called anti-human. The real anti-human thrust involves deliberate acceptance of human demise as a result of merger with machines. If violent approaches were to be adopted to 'solve' the technological problem then it too would become anti-human. Skrbina suggests a non-violent regression of technology, turning the clock back. This is difficult to see. A pragmatic, cosmopolitan approach based of the human person embodied in relationship to nature can work.

Critiques of modernity and technology came both from the left and right. Communists and conservatives shared criticisms but for different reasons. Turning the clock back was more of a conservative wish for some. Both left and right are materialist, use propaganda and want control over us. There is a long line of tech-fear. Mary Shelley (1797-1851) wrote her novel *Frankenstein: The Modern Prometheus* (1818) as another warning. There are a few contexts to consider when we talk about technology and technique.

(a) Technology may refer to technical processes, machines, instruments and devices.

(b) Technology may refer to technique, including the organisation of humankind, knowledge, science and 'science of science.'

(c) Technology may also refer to a combination of the above including any macro-system that 'emerges' therefrom.

Machinery of governance is related to technology. The patent system creates legal machinery that facilitates creation of actual technology. In hi-tech networks, the nature of technology allows strategies of dominance pay off. Firms may use strategies which extend IP rights to extinguish their competitors. So successful have such strategies been that they inform approaches in other spheres. When technology converges through digital networks and more links are made, they become more pervasive. Human input forms a mosaic in a tessellated, fractal and iterated manner to make a bigger system.

Now, we face a growing technosphere that expands daily, almost like the Solid State Entity predicted by John C. Lilly (1915-2001). Humans are becoming things to be managed by machines and technology. Accordingly, we are mancipated or machmancipated. The nature of these networks is to increase through our involvement. Choice can become compulsion. If it was merely being forced to use social media or something else, that might be tolerable. However, the objective is to integrate humans into the machine, technology, system and technosphere. The strange force that seeks this integration of humans is explained in different ways. For some it is merely an accidental, mathematical, economic, technical or commercial function. For others there is a political or ideological reason. For others like Lilly, it may even represent an extra-terrestrial force. Why it is happening is less important than the fact that it is. Real-world evidence is inescapable. The human is becoming firstly dependent on technology and, secondly, is in the process of being merged with the machine. That the future of human evolution is the fusion of human and machine is not merely a fashionable viewpoint. The idea that we would become machines, non-biological and mere parts of technology has been articulated by renowned scientists and scientific writers for over a century. We see it most clearly in the work of J.D. Bernal who influenced Arthur C. Clarke and many others. It is in the work of Julian Huxley who coined the term 'transhumanism.' It echoes in the work of Kurzweil. Scientists, advocates of the Enlightenment and philosophers like Nietzsche (1844-1900), created momentum away from 'normal,' base, slave humanity.

Technology is anti-human in the form it is being produced and promoted. We are being told that we must be sacrificed to technology. There is either massive ignorance or silence. We also face a deadly, total tech-totalitarianism before that. Technology and technique, machinery, 'megamachine' and cybernetics create a new divine network of networks that demand our worship and reverence as a sacred force with its own purpose capable of giving life. The way a society works through their commitment to technology and to invention over tradition means that those no longer beholden will be mastered by technology. The technology will colonise the body, mind and spirit until all that remains are such remnants of the human race that contribute to growth of the technosphere. We are obsolete and must now justify our existence. We are being assumed into the totalitarianism of technique. We face systems based on networks or chains assumed to be always beneficial. The reality is that networks are costly and create opportunity costs or environmental and social ones. These networks form part of the fabric of promissory materialism as indicated by Popper (1902-1994). There are always a set of promises or lures inherent in new technology that may never materialise. What we see is an ever-increasing level of complexity which solves one problem and creates another, whilst creating a degree of anxiety and de-stabilisation that draws the community further into the labyrinth of technosphere topography. While some hope there will be a voluntary unwinding and some seem willing to pursue that goal of regression with aggressive means, we would simply be better to recognise where we are. We cannot adapt intelligently until we understand threats and opportunities.

Chains allow chain-reactions as Leo Szilard (1898-1964) realised when he crossed the road one morning in Russell Square and foresaw consequences for the nuclear era. Near the place where Bernal lived, as did Edgar Allan Poe (1809-1849) who had warned about the 'imp of the perverse,' it was perhaps no surprise. This is one home of magic near the British Museum. Nature allows magic and technology manifests it. But destruction by something 'no bigger than an orange,' as Churchill (1874-1965) had realised, can never be without moral responsibility. While there is always a good reason put forward, there is often a risky drive for curiosity, capability and even celebrity. Scientists cannot stop themselves. The celebration of automatism and robotics is predicated on a redundancy of the human. Marx (1818-1883) and Trotsky (1879-1940) were familiar with Russell Square as was Huxley and Wells. The promise and threat of technology and science became an important part of the discourse of all. In the same Square, spiritualism and psychic research began partly as a reaction to the attempt to dispirit science by the X Club associated with Thomas H. Huxley (1825-1895). They infiltrated other institutions, much like Trotsky would recommend with 'entryism.' It is no longer just the West. If we examine China now, we see a technocratic, surveillance society emanating from the communist foundation using commercial forces to control its system. It is clear that freedom is not a concern, with excessive use of CCTV and a Social Credit System. One might expect that capitalism or the Western world would do better but it tends to the same end. Both systems unfortunately have embraced technocratic control.

International Relations have been primarily based on a 'Realist' idea of relations between states. But power has passed from many states to transnational bodies and most of all corporations. There has been a steady co-ordination of culture internationally as the network of merchandising, marketing and materialism has been spread abroad by corporations who increasingly exert control over state governments and assume governance. Critics of corporate globalism, such as David C. Korten in *When Corporations Rule the World* (1995), have long pointed out how corporate entities create cultural homogeneity in a co-ordinated marketing of a mentality that seeks a culture of longing and aspiration to assist merchandising. Corporate governance created global governance which replaces rights with regulation in favour of large corporate entities. These legal persons grow into the monstrous scientific corporations they were intended to be, if we look back to the writings of J.D. Bernal. It is no mere accident. The Enlightenment always had that capacity to enlighten in the same manner as a pickpocket.

We cannot ignore the phenomenon of transhumanism. This is part of a general takeover by technology moving towards a technosphere. A new artificial environment occurs at the expense of the biosphere and us. An obsession to manage the world means that we too will be managed and mechanised. Such manipulation will continue to all aspects of life if unchecked. Artificial selection will replace natural selection and mechanics will replace biology. In this process of submission, the person is being deconstructed and de-humanised. We have been warned about it by various thinkers. It is inherent in dominant political ideologies. It begins with mass market products that work off networks. The marketing and present Marxeting of such networks and apparent convenience is quickly turned into a vehicle to create dependency and eliminate choice. In all this, science and especially scientism become major forces deconstructing humanity. This is not an English literature discussion. It is not a philosophical endeavour to pass some time on a leafy campus. Rather this book is an effort to indicate the great danger we face. Daft ideas to re-engineer the human through excision, extraction or by taking out supposedly negative features represent a deadly assault on the nature of humanity.

Assault on personal integrity is a necessary instrument for command in the minds of those who administer control systems. Dissolution of human boundaries allows external control via enhancement. Dependency-creation, especially to networks allows shiny chains of control be forged. Dependency on technical links creates connections for control. The individual, human person is deconstructed and reduced. As qualities are being disregarded and the person is regarded as illusory, conditions are being created to increase reliance and integration into networks. As the boundary becomes pierced, sovereignty and free-will will be taken away. The principal import of transhumanism will be to impact on the intrinsic nature of humanity. It will promote the machine, materials and quantity over biology, vitality, bio-viability and qualities. It will take away qualities it underestimates, ignores or is unwilling to explore. The train of transhumanism is freighted with concepts that will drive over humanity. As cyborgs and hybrids proliferate through prosthesis and robotisation, humans will suffer through comparison with mechanical qualities equated with intelligence. Diminished by scientific doctrine we will rapidly decline and our demise described as notional evolution defined by narrow attributes consciousness has been reduced to. Limited thinking is as dangerous as unlimited technology.

The force of technique and technology expands and becomes pervasive, displacing the organism and human activities. Ellul suggests an almost magical force that acts to inevitably oust the organic. Rationality and calculation become dominant through successful production of comfort and things. The relentless logic of technique is to confine humans. It does so through substitution of relations between the person and others and natural environment. Media may refer to the middle, between producers and consumers. Our world is mediated by a membrane of technology. We are gradually assimilated through minds and bodies into an apparatus of authority. This is no longer a metaphorical engagement. We will be assumed physically through lust for a gleaming promise of enhancement through prosthesis. Whether proponents realise or not, transhumanism represents an assault on the human person that could be fatal. As technics, technique and technology grow into the technosphere, the earth is depleted in order to sustain an artificial environment and bio-viability reduced. The technique of transhumanism cannot be ignored in its will to transform the human person based on attacking what it is to be both. Paradoxically, when 'humanism' began, humanists accepted that the universe was inhabited with other beings. Those who attack it often deny that reality yet now seek to create one full of other beings in material form. We are in an inescapable embrace of a hi-tech anaconda.

Scientism and Deconstruction of the Human, Individual Person

"Coming generations will have to take account of this momentous transformation if humanity is not to destroy itself through the might of its own technology and science."

C.G. Jung
The Undiscovered Self (1958)

We may have assumed that we know what the 'human' is. Because the Narcissus of science has become addicted to its own reflection and so fails to perceive the true relationship, it is willing to ignore reality through mistake. We are experiencing an assault on ideas of the human, humanity and personhood representing the greatest gaslighting imaginable. There is a real question about what humanism was or means. The location in Florence of the Platonic Academy was a crucible. Translation of Greek and Hermetic thinking and the growth of de Medici mercantilism, Machiavellian politics and patronage of artistic technique in the context of the Judeo-Christian and Catholic culture suggested a very rich, diverse worldview. Many commentators ignore the wider context in which humanism operated and refer to a desiccated form manifested after The Reformation or Scientific Revolution and Enlightenment. This later humanism had a reduced view of the individual and may be regarded as a type of 'post-humanism' properly so conceived (although it is not labelled so by other people). This disenchanted humanism was a stripping away of wider relationships especially to the Divine. Humanism may be seen as an anthropocentric focus on the individual, reason and experience and factors such as curiosity and creativity. Sometimes humanism associated with the Catholic south of Europe is distinguished from a Northern Protestant type. Some see the core capitalist idea of individualism without institutional mediation as a feature associated with the latter and ideas of the market and panopticon. Humanism was associated with ideas of human dignity but they were the efflorescence of the Greco-Roman, Judeo-Christian, Byzantine and Hermetic cross-pollination. It had many roots and stems. The Enlightenment notion later reduced humanism to an extract, cutting the blossoms and flower from the stem. It focused more on the idea of autonomous, reasonable individuals but those individuals existed in a de-contextualised way, removed from the wider base which humanism originally was rooted in. The Reformation, growth of capitalism, science and Industrial Revolution reduced humanism from comprehensive cognition to people as cogs in the machine. The Catholic Church in the Western world has been reduced and gradually substituted with a comprehensive construct of corporations. Scientism, like the Catholic Church of old, can instead excommunicate, have inquisitions, create heretics, grant indulgences, conquer new dimensions for the new religion or ideology and worship the one true god and trinity of science, commerce and technology.

We are facing a world where technology, technics and governance not only threaten our freedom but will destroy humanity by transforming it into something conditioned by its control. The argument is that transhumanism is liberating and a great promise to humanity. However, transhumanism means technology will determine our evolution and future. Transhumanism and its proponents are willing to sacrifice the human and naturally evolved for some promise based on artificial interference, selection and evolution with the possibility such a trajectory can prevent that evolution science defended so vigorously. Despite the apparently organic and opportunistic development of transhumanism, it really is a tactic in an overall strategy of scientific determinism and control. This is what certain scientists have decided, without any democratic support, will be your future whether you like it or not. They rely on our gullibility, docility and willingness to be enslaved for this to be finalised. It is nearly secured at this stage. In looking to the dangers of an anti-human, de-personalising force we must engage in consideration of what a human or a person is? We must also remember that as technology increases, humanity may decrease. There is transference of environmental and social capital into tech-control. Google, Apple, Tesla, SpaceX, Silicon Valley, Bostrom, Esfandiary or FM-2030 (1930-2000), Ettinger (1918-2011), Kurzweil, Julian Huxley, Musk, Max More, Moravec, Thiel and others combined with Chinese and Russian developments means that this agenda has come from an apparently fringe position, sometimes libertarian, into the vanguard of hi-tech development. However the supposed libertarian ethos is refuted by the actual historical path back to J.D. Bernal in 1929. By the time Teilhard de Chardin (1881-1955) anticipated interaction of machines at a global level and the combining of consciousness, he was almost providing a theology for transhumanism.

Teilhard had that same almost pathological need for certainty. Bernal and Teilhard require crystallisation of society so it is more predictable. Teilhard describes a lotus shape emerging from the mass of people.

> *"At the heart of this huge calyx, beneath the pressure of its in-folding, a centre of power has been revealed where spiritual energy, gradually released by a vast totalitarian mechanism, then concentrated by heredity within a sort of super-brain, has little by little been transformed into a common vision growing ever more intense."*
>
> The Future of Man (1964)

Opposed to Teilhard is the stream of thinkers that emphasise the person. Thinkers like Scheler (1874-1928) suggested a philosophy and ethics based on the whole person. He allowed the internal apprehension of higher values and higher feelings beyond interactivity. This was not rejection of reason but a shifting away from undue reliance on Kantian reason. He suggested a place of objectivity of feeling beyond that. There is often a long list of philosophers like Bacon (1561-1626), Newton (1642-1727), Hobbes (1588-1679), Locke (1632-1704), Darwin (1809-1882), Bentham (1748-1832) and Mill (1806-1873) to contend with in transhumanist discourse. It is important to make sure one is on firm ground before one does. Addressing these issues is necessary. One idea that is useful is the idea of 'technique.' Remember also that reduction of humanism and negative associations therewith was done by forces which now claim to cure it.

Ellul in *The Technological Society* argued that 'technique' in general obeys its own law and will be heightened by mechanical technique to an incalculable degree. At page 12 he wrote,

> *"the ideal for which technique strives is the mechanization of everything it encounters."*

It was clear for him that humans were in the line of fire. He indicated the momentum within technology.

> *"Technology has become autonomous; it has fashioned an omnivorous world which obeys its own laws and which has renounced all tradition."*

He thought that psychoanalysis and sociology had passed into technique. Method then takes over. At page 18,

> *"The "scientific" position frequently consists of denying the existence of whatever does not belong to the current scientific method."*

Then technique is reduced to numbers. Technique could be anything. When efficiency is the sole goal and action is standardised or reduced it becomes something else. Then technique is extended. Numbers determine approaches. Specialists decide. Calculations are king. Technique includes economic, human and organisational methods. Ellul thought that magic was the first, primitive technique of the spiritual order related to human's efforts to control, although this has been overlooked because of more material-orientated technique. So there was a link between worship of technique and magic. He thought the Greeks developed a theoretical, abstract science divorced from technique. Science was more related to wisdom. The Romans developed social and legal technique. The French Revolution and Napoleon, in particular, sought to have pervasive technique and control through quantification and organisation. The nineteenth century involved a closer relation of science to technique as the former began to serve the latter until the twentieth century. At page 45 he writes,

> *"...this resulted in the enslavement of science to technique."*

Philosophy pushed science into this relationship through utilitarianism, empiricism and a focus on the material and practical applications. He argues that groups were disbanded to atomise society notionally in favour of the individual, which led to a type of slavery. In Britain, he uses Cromwell as a starting point. Later in France the State became powerful. Industry encouraged invention from 1750-1850 to develop efficiency. Social systems were broken down by industry and enclosure. Industry, commerce and technique inform the State reflexively.

At the end of his long exploration Ellul is shocked by the banality of scientists outside their range of knowledge and their willingness to take control, make us happy and manipulate our perceptions. Ultimately he concludes at page 434,

> *"That it is to be a dictatorship of test tubes rather than of hobnailed boots will not make it any less a dictatorship."*

The result of technique will be 'a world-wide totalitarian dictatorship.' He argues that we are being dismembered to serve technique. Transhumanism demonstrates that he was right. He saw it in the French context in particular. British scientists told us what they were going to do. Wells explained it. If after recent events and loss of liberty combined with a great advance in technology around us, you cannot accept that which scientists like Bernal have said would happen is happening, then it may be because you have skin in the game or are in on the game or think you will win the game or else you have buried your head in the sand of fine grains of misinformation, models, authority figures, statistics and prognostication all falling in the time glass. With our very nervous system linked to algorithms that manage us, we are yielding finally as a race, sinking in the quicksand. We have wandered into a marsh following the will o' the wisps of images that led our fantasies and the heart has indeed grown brutal with the fare as we succumb to fear. While we look to norms to settle these ethical issues we see that they are eroded by the forces we seek to examine as Hans Jonas noted in 'Technology and Responsibility: Reflections on the New Tasks of Ethics' (1972). Ironically, while we will suffer de-personalisation, corporations or legal persons gain rights at our expense. Robots and other beings will also gain rights as we lose them. When you have relinquished sovereignty to elite groups of technocrats, you can expect them to re-engineer the game to suit their panopoly. You will soon see that the technopoly has a panopoly of machines primarily calculated to manage us. You. Us. It has already been shown how we can have our homes turned into a workhouse with figures and models. As machines are added, we will also lose. Many machines and techniques will subtract something vital from us. The relationship between subtraction from us and replacement or addition through mechanisation or silicon is part of the process of prosthesis. As things are added to us, we will lose things and qualities, some clear some subtle. We lose that which is unquantifiable. We will even lose our humanity. This could not occur without some degree of an anti-human bias or misanthropy. Unfortunately, the benefits of absolute control of other humans are another great incentive to some. Some claim transhumanism is a product of the Enlightenment notion of reason, progress and technology, some reject this. Posthumanism is more related to postmodernism and ideas of deconstruction. Such ideas supposedly do not advocate progress and reason in the same way. The origin of the term is located in an article entitled 'Prometheus as Performer: Towards a Posthumanist Culture' (1977). The link with Prometheus is worth noting. This is now a widely used term with diverse meanings. Transhumanism can appear similarly fragmented into libertarian, 'extropian' or democratic types. Anti-humanism has specific manifestations such as the Marxist type. I have identified my definition and will indicate more specifically what posthumanism is. All represent an attack on the human. It is clear to thinkers like Francesca Ferrando that science and technology have a big role in fragmentation of the idea of the human. Posthumanism may be a later stage transhumanism or it may be a separate deconstruction for other purposes such as feminism. It is likely that both are merely different ways around a roundabout with technology being the determining destructive force. Even some feminist proponents of transhumanism have found that their power may be used to promote goals inconsistent with their original aims. This reinforces the idea that science with a hi-tech impetus is the driving force in transhumanism. Other debates are complex but ultimately academic. Science fundamentally made or caused transhumanism and posthumanism, whatever flavours they may have. Much posthumanism claims that because many humans were not regarded as humans thus humanism was flawed. This seems often to be an ideological argument.

The industrial model of processing animals for slaughter has informed our very thinking. Deconstruction is consistent with the assembly-line that takes a living thing apart and reduces it to constituent elements. In that process of reduction, the vital force is taken away. Procrustean process controls all by taking something away or alternatively adding. Science often misses the vital force of consciousness which makes science itself. The human, individual person is being taken apart and enslaved in hi-tech networks until a shiny mechanical simulacrum can replace it. Thus managed, biological 'machines' will diminish in value while their artificial substitutes will be valued and accorded more rights than the mechanising agents, formerly humans, who are being transitioned away from their messy personhood. This is not my conspiratorial imagining but implications from what scientists and their surrogates have said. We are in an Empire of Scientism, a scientocracy where others feel entitled to manage us totally. It is not merely an issue of limitations of science or scientists as espoused for example in *Science is a Sacred Cow* (1950) by Anthony Standen (1907-1993). Worse now it is an issue of a deadly belief in the religion of science, the ideology of scientism and actual elevation of the technocratic, machine to displacement of others and even people in an apparatus. Scientism largely inspires other philosophical forces of deconstruction. An insane confusion of the product with the producer creates a type of Oedipus complex which glorifies one parent and wants to kill the other. We have been beguiled and bewildered so much that we believe the fruit of invention, investigation and innovation are greater than the plant from which it was necessarily borne. 'Hubris,' in the sense of excessive pride, in this case resulting from power of huge success, has given rise to 'hubris' in its original Greek meaning, which represented a violent degradation or humiliation. The persistent desire for oligarchical control of other people will utilise this philosophy and practice to reduce humanity. A pragmatic assessment of transhumanist policy indicates that the promise will turn to an imprisoning purpose and supposed nobility of motivation is often mere opportunism masquerading as majestic enhancement. It is not the individuals with dreams that should worry us but possibilities for mass application through compulsion by old, reliable, imperial tendencies in human affairs as informed by whatever principalities they happen to believe in. The Empire of Scientism succeeds through combination of different legal and corporate personality to derive transnational power to behave independently. Ironically such real, technical power is based on denial of inherent force of the whole, individual organism in biology and a sacrifice of the human, physical person to the real, legal fiction and power of corporate personality. It rejects the divine and especially a singular Divine.

Scientism as ideology seeks then to promote science and technology as the inevitable solution to all problems. Transhumanism creates a story that is supported by scientism because it is generally a manifestation thereof. This may be just a mistake or a deliberate policy. In its effort to promote desirability of reconstruction of the human it inevitably needs to deconstruct them. The deconstruction and associated de-humanisation is also inherent in the pressure of technique and associated centralising and converging force of global governance. Totalitarianism or tyranny requires de-humanisation and often the body of the subject turned into an object. The dishonesty in transhumanism and posthumanism is a manifestation of the reduction of humanism by science, reason and even Reformation. Early humanism was based on a much wider basis that was potentially more holistic and cosmopolitan. The dispiriting forces of the Industrial Revolution also contributed to a mechanistic view of the human. Granted the Church was complicit in negative forces and events and acted as an Empire or global actor. However, the continuing dynamo of the focus on rationality and creation of instruments ironically was also involved in these forces of conquest and the growth of European imperialism. Now we are to believe that the mistakes of previous empires are to be cured by creating a new version that unites all existing one or excises competitors. This new Empire will not only control people but it will literally rule their minds and bodies because it can now do so technically through implantations. Transhumanism is based on a material, materialist misrepresentation of what humanism was and thus tilts at windmills whilst ignoring the compost of its own origins.

The Renaissance, particularly in Florence, is the source of much of our conception of humanism. Humanism rediscovered or underlined the central focus of the human as measure of all things and microcosm of the macrocosm. Art, architecture and science focused on the individual human person in the context of an accepted worldview presuming a divine element. But there were two distinct trends in my opinion. One was holistic and saw the individual as embedded in a multiverse from where humans could evolve to be higher beings or regress. Knowledge could be sought anywhere from any tradition and involved recognition of spirituality and humankind that was universal and respectful of variety. Another trend was to solely concentrate on the power of the rational as expressed in mathematics and manifestation in the physical or material world and tend towards science. This was most obvious in the work of Brunelleschi (1377-1446) in the construction of the Duomo, use of optical devices in painting and development of perspective as a mathematical solution that effectively developed illusion. Illusion is now argued to be inherent in our perception by contemporary scientists and used to belittle our perception and abilities although the science of art was a masterpiece in creation of illusion. Perspective is an illusion, a trick to reproduce a 3D scene in 2D. The growth of banking and business was also associated with this time of abstraction and complex construction of symbolic worlds. If you take particular aspects of this period and condition them falsely on forces that are seen to be regressive today, then you can continue the Enlightenment disintegration of all humans oblivious to the irony that you claim to improve humankind by interfering. Much later, 'humanism' was narrowed to mean a conception of humans without religion and it became defined by its denial of God. This more contemporary dimension of humanism becomes a more materialist, physicalist view of the world and may celebrate the secular mundane. Secular, modern, scientific, naturalistic, ethical humanism, where it is against religion, supernaturalism and any other divine context, represents something other than early humanism. It may be the culmination of one trend perhaps from the Renaissance but may also be more informed by the process of Enlightenment dispiriting. It is a mistake to project this contemporary viewpoint back through history and presume that it has much relation with the early origins of humanism. Straw humanism might be used to refer to the establishment of a 'straw man' argument in relation to humanism. By taking an artificial idea of humanism and burdening that plastic mule with a lot of unrelated flotsam and jetsam of historical injustice, eventually you will get a Buckaroo effect which jettisons all to allow alternative conceptual caparison. Contemporary humanism has many admirable qualities (and some of my best friends are humanists) but in its association with certain figures like Julian Huxley the objective seems clear. Many strands in humanism are 'scientific' and seek to create a collage of meaning to satisfy needs it will not otherwise accept. Deep spiritual longing in humanity cannot be ignored. The clearest recognition of the sandy base of such humanism is that it leads to transhumanism and posthumanism. If the human is to be celebrated by humanism which is scientific or modern, why do those proponents want to transform us? They forget that to look at the being we must realise first that we are actually standing on the ground of being.

The human person is being deconstructed by science and more particularly scientism and through associated philosophical forces often predicated on materialism or collectivism. The technique that deconstructs in this way is not liberating but instead part of the illiberal trend to scientocracy explained and advocated by scientists. In posing a biased and narrow conception of humanism, they have rejected its true holistic base. They have confused their notional human construct and the actual human person. In addition, they may obscure the distinction between philosophical or theological categories and mundane legal discrimination on the basis of legal recognition of personality. Indeed, it was people like the Dominican Friar Bartolomé de las Casas (1484-1566) who argued that indigenous people were human because of his theology against others who denied it. While this is not much balm for the horrific, assault on the Americas, it is noteworthy that the presumption of personhood is the base of humanity and the fuller sense of humanism that existed, in contradistinction to the colonial lust for things, treasure, plunder and destruction. Now we are to presume that the lust for things and even the possibility of transforming a person into a thing, driven by a similar acquisitive mind that sees nature and her people as objects to be controlled and exploited, is the superior attitude. The new Empire is a continuance of acquisitive forces of colonialism driven by technique and technology to colonise the individual, human person through breaking conceptual boundaries by abstraction to penetrate their being and de-humanise them in the disguise of a new era of enlightened 'civilisation.' Without violence and vilification we must defend ourselves or otherwise become the supreme victims.

Imperial Incentive to Transhumanism: All Roads Lead to Rome

> *"...if you think that if I become a governor the devil will carry off my soul, I'd prefer to go to heaven as Sancho than to hell as a governor."*
>
> *Don Quixote* (1605) Cervantes

My pamphlet *Empire of Scientism* outlined emergence of an imperial structure of scientocracy which would become totalitarian. The process and tricks of transhumanism suit the agenda of the emergent Empire. The immediate benefit of transhumanism is that the individual may become dependent on some other material or process that extends the technosphere. The greater a technological dimension the lesser the relative existing, biological functionality. Any way humans can become dependent is suitable for expansion of the technosphere and Empire of Scientism. Elon Musk says that we are cyborgs because of mobiles. If they evolve to implants or attachments we have techno-enhancement consistent with transhumanism. Like the mobile became akin to an electronic tag, so the implant would make a system of control. If we are made dependent on networks, the desire of controllers to create network effects implies possibility of greater control through enhancements. In a technocracy or scientocracy, incentive to use transhumanism as a method of totalitarian control is very high. Such techno-totalitarianism would not be a great orchestration of enhancement for the population motivated by benevolent policies but employment of the lowest common denominator of intervention for control by technological compulsion.

The process of normalisation of new hi-tech usually focuses on obvious benefits that such technology brings. Marketing and manipulation of public opinion will focus on apparent advantages obviously. Social costs are seldom addressed. The price we pay is forgotten or disregarded. There is a presumption of superiority of all technology and machines. We like toys and objects in a fetishistic way. The process of tech-network-dependency becomes normal. Dependency proceeds then through a narrative of unavoidability. This procedure of compulsion is merely to prepare us for further integration. There is no easy way back. Beneficiaries of power-dividends from network control are not benevolent. Controllers of the new Empire of Scientism have power to decide to sacrifice humanity or the vast bulk of it for its own ends. Some have assumed an obligation to exert their desire to control by technology the future of the world, in a way similar to how they already command people's conduct. The behaviourist impulse and a desire to cut out negative features of humanity, and eventually free will, can be enforced on the global population. Mis-sold as objectives determined by supposed necessity, managers and manipulators who assume power from their own will, enhanced through combination and made possible by machines which suit their materialist and mechanistic worldview may maximise their control and domination at our expense. It is us humans that will be engineered by engineers.

The saying 'All roads lead to Rome' indicated a truth. An imperial structure needs to be at the centre of a system of communications to march its forces and enforce control as rapidly as possible. Studies have indicated how the road system in Europe does actually funnel towards Rome. Many of the major highway systems are still based on what were Roman networks. Philip K. Dick (1929-1982) believed that the Roman Empire never really went away. In my novel *Blue Lies September* (2019), a character considers such a proposition on the basis of similarities between the conduct of life in Roman times and the present. He did this on seeing contemporary emphasis on trade, tax, the military-industry complex, the use of slaves and tactic of entertainment for the docile and powerless plebeian class.

The concept of *translatio imperii* is a suggestion that power is transferred from one imperial structure to another. It is like a principle of imperial energy which implies that empires are not destroyed but transmute from one form into another. Thus imperial tendency recurrent throughout history transmutes into optimum form for survival. Hi-tech networks associated with bio-security are the ideal vector for the ultimate imperial infrastructure. All roads do not lead to Rome, but the information superhighways lead somewhere. We know now that they lead to commercial centres that are congregated and security networks in the US and elsewhere. The link between the scientific corporations and state apparatus of the US, China, UK, Russia and others beyond effective reach of the citizen, means that the nucleus of governance through despotic elements, can coalesce round objectives that are shared, such as scientism and other life-or-death justifications concocted thereby to ensure a cohering among the cadres likely to benefit from emergence of the new ruling class of the emergent Empire.

The recent global crisis represents the shift away from the tactic of globalisation to the reality of the new world order based on rule by experts and scientists without democratic or religious restraints. The infrastructure of global governance through bio-info tech means that the pretence of globalisation can be abandoned, as the charade it was, to accelerate deconstruction of the nation state. Sources of homogeneity consistent with a historical tendency to assert anti-imperial sentiment needed to be made more heterogeneous so that the possibility of opposition could be reduced. An inchoate apparatus of imperial scientocracy of a totalitarian tendency is in place. Bearing in mind that this illustrates predictions and policies articulated openly by the scientific community a century ago, why disbelieve scientists? Assuming that the agenda proposed by them has come true, it is logical, reasonable and rational to conclude that the other aspects will come to fruition. It was a central part of the agenda as expressed by J.D. Bernal that the human body would be mechanised and the docile population would be joined up in some way. This transhumanism agenda is not a fantasy or the produce of some paranoid mind but the actual desire of scientists (and their sycophants) who are still respected and who have influenced the 'science of science' or infrastructure of the world. Thus transhumanism will be inflicted on us. We will have no choice. It will be an objective that appears with an apparatus to enforce it, as scientists have said it would. Transhumanism ensures total tech-totalitarianism. Cyberspace assumes and incorporates us as cyborgs and then assimilates and consumes us.

Certain people sensed where this was going such as Kafka (1883-1924), Arendt (1906-1975), Anders (1902-1992), C.S. Lewis, Ellul, Charbonneau and Philip K. Dick. This was the way Bill Joy put it in his essay on 'Why the Future Doesn't Need Us.'

> "...but control over large systems of machines will be in the hands of a tiny elite- just as it is today, but with two differences. Due to improved techniques the elite will have greater control over the masses; and because human work will no longer be necessary the masses will be superfluous, a useless burden on the system. If the elite is ruthless they may simply decide to exterminate the mass of humanity. If they are humane they may use propaganda or other psychological or biological techniques to reduce the birth rate until the mass of humanity becomes extinct, leaving the world to the elite..."

The most logical way to sell order is to create disorder first. If chaos is unleashed, calls for control will increase. The Western world is encountering a decline in order within nations and an increase in violence despite much massaging of figures as appropriate for whatever purpose presents. The substratum of society is under pressure. If disorder increases then eventually order will be sought or imposed. Increasing disorder is associated with a narrative adumbrated by H.G. Wells and others in the blueprints and 'open conspiracy' for a 'new world order' that he advocated nearly a century ago. In order to attain global governance by scientific elite it is necessary to eradicate 'false loyalties.' False loyalties represent the social glue that creates coherence. We have been convinced to deconstruct society in order for it to be reconstructed or 'built back better.' Re-construction will involve a vast network of a controlling, surveillance apparatus based on an unholy alliance between bio-medical control and IT companies. Science in its ethical neutrality and moral cowardice will cause crystallisation in conjunction with the ideology of scientism. This is observable. See if your freedom is increasing or decreasing. Your power is relinquished to someone else. The new Rome will have such actual power that the old Rome at its height will look like a group of scouts at Sunday School. The vague idea that some freedom-loving hackers and moral activists are able to counteract this emergent structure is deluded. A straitjacket of tech-bondage will compromise any activity. Martin Luther King (1929-1968), Gandhi (1869-1948) and Daniel O'Connell (1775-1847) were brave people who achieved much through peaceful agitation and activism, often in the face of deadly hostility. Arguably they were operating in contexts where there were more civil rights such as freedom of assembly than at the time of writing. One can see this as a temporary context, analogous to a war. Nevertheless there is a strong probability that a precedent has been marking a shift.

The shift is to global governance driven by scientism. Scientists for the last century, and before that in Russia, have indicated the desirability of transforming the body by non-biological material. Whether part of this thrust or not, the likelihood of transhumanism increases. The greatest probability of transhumanism occurring is through compulsion in order to control people. This justification may spuriously be posited on the basis of protecting the environment or public health or something else. Whether mass transhumanism, beyond reasonable expectations of restorative medicine, may be driven by mad scientists, libertarian entrepreneurs, surrogates of scientism, useful idiots or genuinely open-minded explorers, the danger of using such technology primarily as an instrument of global governance is very real. Such transhumanism at the start will be transhumanism-lite and may involve some small implants. Transhumanism-lite will be on a mass scale. This initially seems to come from the nature of networks. When the telephone was invented, some enlightened early adopting Mayor in the US foresaw the day when every city would have one. That is not really how it works. Manufacturers want to make one each, maybe two or three. A unit in each home or office makes economic sense. This derives from economies of scale. The necklace of networks allows dependency wherewith we must now integrate, perhaps for work or school or communication. Like E.M. Forster (1879-1970) predicted in his 1909 work 'The Machine Stops' we will be controlled by 'The Machine.' The next step is clear. When you have forced people to be at home, atomised and controlled by network technology, then total governance is easy. Noble benefits may be cited. The need to protect the environment, to attack overpopulation to ensure order, to stop crime, to manage resources or to deal with the latest 'crisis' can act as cohering forces in those concentrations of power that have the means and will to govern us globally. We are on the threshold of the dream of turning humans into machines. The first step is to merge with them. Like the journey of a thousand miles starts with one step, or a change of condition with one vampire bite, it only needs one relinquishing of physiological sovereignty to the Machine for transhumanism to occur. The likelihood however is that it would occur at a rapid pace. If a bridgehead is made into our consciousness we will be managed by AI in the network, more directly than now, through machine mesmerism. Associated with the Empire of Scientism, expertise, managerialism, a cult of codes, calculation and ratiocination, is a disposition which seems to dislike humans or even hate them. The love one might expect for one's fellows is transformed into a desire for machines as a manifestation of self-love or a retreat from the unpredictability of a quantum world into a materialist, imperial system. While there are people who have a pathological love for power, there are others who will support the apparatus because of a discontent with humanity. I label this disposition 'machine misanthropy.' In trance, tricked through succumbing to base instincts, we may allow ourselves be subjected to reduction and supposed augmentation. That addition is a poisoned fruit. Confounding lubricated compulsion with choice, allowing technocrats take remote-control of us through our super-max greed or fear, we end up in a prison. We must recognise that a certain type of misanthropy informs this obsessive drive towards the hi-tech zoo. The unique technophile policy derives from and promotes a certain psychological disposition that will prosper. Even still, corporate personality and the blind force of technique do not need bad actors in order to work. It is possible that some transhumanists do not realise that they are pawns.

Psychopathocracy or Machine Misanthropy

"Homer, I'm afraid this is the part of God's perfect plan where you're murdered by robots."

Ned Flanders in *The Simpsons*

Materialism may metastasise into a deadly mind-set. As I sought to indicate in the *Empire of Scientism*, there are a number of forces contributing to the vortex of tech-totalitarianism. There is simple opportunism deriving from possibilities of power provided by the process. To some extent it is a product of the mentality that wants to create technical order and is inherent in the technology itself. Technology has a life of its own. There is a force of drift. Machinery of machinery, especially with AI, has capacity to self-organise and determine our lives. Allied to this, technology could not happen without humans. We are knowingly giving birth to this monster which we then claim is beyond our control. It is constantly suggested that there is an inherent, exponential essence in computing potential that must result in our enslavement and demise. This is merely another version of the idea of science and technology as magic or more like black magic insofar as it may represent species-suicide. It is difficult to explain accelerationism behind technology as advocated by certain proponents. It could be regarded as something almost psychopathic, Faustian or some other type of kamikaze death-wish at times, perhaps driven by disenchantment.

Imagine we were to be run by psychopaths. In many senses psychopaths make good leaders. They seem to be over-represented in high-level management. They may demonstrate a quick ability to spot and exploit personal weaknesses. They have shallow affect. They may manage their image well. They will have less restraint than others with a conscience. They fit well into a machine for implementing control systems. Con and troll. The qualities of superficial charm, shallow affect and ability to exploit people around them sound like something robotic. If an enterprise or system seeks to achieve power to promote some ideology, then psychopaths may be very useful and fit into the empire-building stage very nicely. People who have grandiose plans to rule the universe are probably in some similar category. The ideology of scientism has a certain psychopathic feel about it, if one looks at the literature and compares characteristics with the persona of the movement and individuals in the vanguard. If we were in the phase before final implementation of a global governance dictatorship what do you think it would look like? Despots and crackpots are always reasonable or rational to some degree and will need to create the impression of concern for the common good at times. When power has been gained, the mask will fall. Bernal in his book *The World, The Flesh and the Devil: An Enquiry into the Future of the Three Enemies of the Rational Soul* (1929) suggested that scientific corporations could gain power by stealth and thereafter unveil the truth. That leading scientists wrote openly about the enterprise to seize power is remarkable. Wells even called it *The Open Conspiracy* (1928).

Hi-tech billionaires have been able to lever symbolic logic dispositions to use public-granted monopoly and exclusive rights to build business empires. Techniques that worked in industry are replicated into a broader notion of tech-governance. Cybernetics is about governance. It is perhaps unsurprising that those who have attained great success and power through applying their mode of thinking and projecting it then believe that their success should be applied to other issues. Everything begins to look like a nail for corporate computer hammers. When we engage with the computer-matrix, we are entering into a manifestation of a type of tech-thinking of persons embedded in technique. They project the power of their mind onto others with technique. When such a person sees technical, economic and commercial growth of ideas implemented into machines they may think it appropriate to advance more of the insights they have obviously given in the machine's embodied form. Ubiquity of their inventions, innovations and power deriving from technical success must make it tempting to apply those successful ideas to machinery of governance. The technosphere grows to accommodate success of technique and technology with many inputs and few limitations. Military technique seems to ultimately be re-applied by industry to control the citizens who funded them. I do not criticise successful people but how pervasive techniques of control, through technology uncontrolled by adequate application of law, can lead to possibilities of command by others who utilise computers and networks. Prometheus is close to Faust. The story suggests the danger of stealing technology or fooling the gods yet it is turned into a story of triumph not tragedy by those who use it.

An example of anticipation of this society can be seen in the 'The Machine Stops' by E.M. Forster. He sees the dark potential of the underground, atomised society of weak humans with skin white as fungus. That is the fear or prediction. Today, if one mentions a concern you are called fearful, acquiesce you are courageous. Inversion is a tactic. Forster's warning might be contrasted with the optimism in the work of Fyodorov (1829-1903). This Russian philosopher was a forerunner of transhumanism. He anticipated mechanisation of the human, imagined practical resurrection and enhancement of weak biological systems. If you accept some of his assumptions and think through the consequences, it becomes clear that they may be workable in specific and limited contexts, but that massive commitment thereto would guarantee a hell rather than a heaven. Forster's picture seems more accurate. Russian 'Cosmism,' imbued with a tech-utopianism, promises much more than it will deliver. Zamyatin (1884-1937) seems to support Forster's concerns where personal isolation and manipulation, or what we might call humanipulation, is the norm. Humans are constrained by the web they weave as an artificial garment. This may become a magician's cloak. Forster anticipated in The Machine that the technosphere would perpetuate itself and that humans would become subservient to it. The hope cherished would be that its focus on goals of maintaining itself and operating would make it unable to adapt.

The movement of Futurism in the early twentieth century, which informed Fascism, was predicated on an accelerationist theory which sought to glorify the machine, speed and war. There seems to be a strange kamikaze spirit linked to technology. This is not only the nature of

Icarus who flew too close to the sun, but also an inherent idea of destruction to oppose creation. If we create an infrastructure of industrial development that attracts those who will take risks with others and the environment on the one hand while creating an excessively restrictive regime for the ordinary masses on the other, without opportunity for democratic control as a result of a shift of power to transnational networks, then we have an apparatus for an authoritarian scientocracy. Local accountability dwindles. The idea of a collection of rational individuals who do not share a philosophical base and have discarded moral systems as superstitions may create conditions for something like a psychopathocracy. But we do not need to consider the extremes to identify a unifying idea. A proper sense and use of the concept of 'misanthropy' clarifies. This term that comes from Greek refers to a hatred of humanity. It has been softened to mean something lesser. But let us take misanthropy as meaning a genuine hatred of humanity. In that sense misanthropy is unfortunately an accurate description of the impacts of much governance. The reason why so many bad decisions are made is because of such self-hatred in humankind. Misanthropy may be at the basis of much of the transhumanist and posthumanist endeavour. Why else would people want to end humanity?

The idea that some governors hate people seems preposterous in one sense. However, when you match it to the list of despots and warmongers we have had, even in democracies, it is clear that misanthropy is a sound explanation. It is easier to promote massacre if you hate people. While Hitler proclaimed his love of the German people, he was also willing to destroy them. Misanthropy is often linked with a love of machines. Machines can be programmed and do what you want. They can exercise your will faithfully. They will not criticise you (yet). They give you power over other people. They can reassure their makers that there is a plan and a way to make the world around them predictable and not fuzzy nor messy nor requiring of human interaction. They can greatly reduce population in wars. People can have sex with machines rather than people. They can move in a car rather than walk. Perhaps if you behave more like machines, the people who manage machines can accept you more. Perhaps if you behave more mechanistically you will be more predictable and fit into Procrustean places prepared by the policy-makers for you. We will increasingly be engineered to fit into the technosphere. This only requires will and acquiescence. Humanity is to be re-conceived, re-defined, controlled, rendered predictable, altered, reduced, minimised and mechanised. Meanwhile machines are maximised. Misanthropes love machines that help destroy other people and control them. I suggest that Machine Misanthropy is a real phenomenon. *Machine Misanthropy may refer to an actual or intellectual hatred of humanity and a consequent or associated, projected love of machines combined with the tendency to use it as a machinery of control to replace or remake the human.* One need only look at how an expensive car may impress certain people to the detriment of other people to deduce how a mechanically-enhanced body may do the same. Some Ferrari-level technological addition with some practical benefits might create enhancement by wonder and desire and through status elevation could foment the popularity of the phenomenon itself to foster competition.

The motive force behind transhumanism is presented as a policy of addition whilst in reality it is predicated on a force of subtraction. In the guise of getting something you will lose and have things taken away. Loss occurs through the reductive view of humans which denies personhood and qualities associated with consciousness that are not mere computation. An additional loss is through ceding control or sovereignty to systems of technical intervention and costs of dependency. You will be told that integrity and boundaries are bad. Privacy is bad. Colonisation of the human body requires an attack on barriers of personal sovereignty. Colonisers must attack boundaries. Spurious justifications and supposed benefits promote inroads into human consciousness by artificial computation and algorithms. There is no need to construct some conspiracy to colonise the human body or mind. This is because scientists have told us that this is a desirable destination of scientific endeavour. Even if people wish to ignore what scientists have said, it is clear that there is irrefutable evidence of the tendency of technique and technology to centralising control. Transhumanism as a tool in the hands of commerce and institutional actors that seek governance may be wielded against the individual despite contrary claims that it is calculated to promote individual freedom. The bridgehead in our body and head is via prosthesis. Techniques of persuasion are many more when there is no moral base. As Banquo said in relation to the prophecy in *Macbeth* (ca.1606) the instruments of darkness win us to our harm telling truths about trifles and then betraying us in matters of deepest consequence. The most dangerous can manipulate us because of our trust and expectations of fair play. Gaslighting and inversion are everywhere.

We know the cure often comes from where the poison is. It is a mistake to ignore the disciplines or domains that seem to promote the pervasiveness of technology that may make the simple complex and remove us more by compelling face or contact time with the technosphere. We will become so dependent that our ability to escape it will diminish. When that ability is so reduced, our ability to withstand incorporation thereto will disappear and network transhumanism will follow. Stafford Beer (1926-2002) was a theorist very interested in cybernetics and management. He criticised capitalism and communism. He recognised that both suffer from 'dysfunctional over-centrality.' Instead, he explained that there should be maximum autonomy in the parts of a system which do not jeopardise the system integrity. Autonomy is the degree of freedom consistent with the ability of the system to function. Those seeking global control seek centralisation. Apart from the will to power, the system of administration requires adherence to certain rules. It may be that there is confidence somewhere that the knowledge and practice of contemporary computers and models of conscious behaviour or agency create a belief in an opportunity for centrality of control and an associated belief in an obligation to do it for perceived global welfare. However the loss of variety in the drive to centrality will create great flaws which usually invite more of the same by those convinced in the centralising force. Cybernetics suggests dynamic systems but the use thereof has tended towards centralised systems coalescing around the personal belief of key players in momentum to a global tech-eschaton. Systems have their own rules that may make people behave in a particular way independent of their will.

Prosthesis of Transhumanism to Posthumanist Prison

"The doors of heaven and hell are adjacent and identical."

Nikos Kazantzakis
The Last Temptation of Christ (1955)

The promise of something that can be added to you, to improve your life, is the essence of the consumerist dream. Inculcation of disenchantment and a bait or lure of enhancement to solve implanted discontent is a ubiquitous strategy to inoculate against spiritual sovereignty. Easy superpowers are suggested without need to cultivate siddhis. We are in trance, entranced by transhumanism at the entrance to a new era. Entrance is the exit from humanity as we know it. People assume that technology just happens. More knowledgeable people will understand how the patent system plays a critical role, as do the institutions of science. Few remember that technology is a crucial part of other ideologies as well as being part of an ideological approach itself. Technology formed part of scientific socialism or communism. Within that narrative, there was a strong sense of the way 'new man' would be improved by machinery of the system. Scientific socialists like Bernal openly explained inevitable transmutation of humankind into technology. From the 1920's, all around the world, the idea of a future utopia inspired by science was advanced by many significant thinkers. H.G. Wells indicated how far the scientific socialist movement should go in *The Open Conspiracy* (1928) and he wrote,

> *"The cosmopolitan revolution to a world collectivism, which is the only alternative to chaos and degeneration before mankind, has to go much further than the Russian; it has to be more thorough and better conceived and its achievements demands a much more heroic and more steadfast thrust."*
>
> H.G. Wells
> *The New World Order* (1940)

One of the leaders of that Revolution also indicated the glorious future helped by technology. Trotsky spoke in similar vein,

> *"More correctly, the shell in which the cultural construction and self-education of Communist man will be enclosed, will develop all the vital elements of contemporary art to the highest point. Man will become immeasurably stronger, wiser and subtler; his body will become more harmonized, his movements more rhythmic, his voice more musical. The forms of life will become dynamically dramatic. The average human type will rise to the heights of an Aristotle, a Goethe, or a Marx. And above this ridge new peaks will rise."*
>
> Leon Trotsky
> *'Socialism Will Bring Great Advances for Mankind'* (1924)

Apart from communism, which built on the British, French and Russian tradition of machine-humans, there is a possibility that transhumanism will be a spearhead in the establishment of a system of global governance uniting communism and capitalism. My two previous books, *Empire of Scientism* and *TechBondAge*, lay out the following arguments. We are facing a totalitarian Empire based on governance of converging technologies. The Empire is to issue an era I called the 'TechBondAge' with techbondage describing the method of enslavement on a global level. This was anticipated by science and identified by tech-critics. Günther Anders is one writer who perceived implications of technology through examining tv and its impact. But in addition, the nature of technology and technique create inevitable and inherent momentum to a global totalitarian system of control that will involve transhumanism. So transhumanism can come as a result of ideology, the inherent nature of technology or through the free market or capitalist system. In the latter case there are a few distinct possibilities. If it comes from individuals exploring their own nature with their own resources or if associated with medical restoration then it is less of a concern. However, the danger comes from emanations based in the security, military-industrial or pharmaceutical complex. Converged power available through networks and congealed in transnational spaces beyond national regulation creates a possibility or a probability of imposing transhumanism on some pretext for profit. The fact that transhumanism may be imposed upon us is a matter of grave concern. While people will see cyborgs before them and conclude that it is very unlikely that we will be transformed any time soon into any such machine, we must understand that it only needs an apparently minimalist implant into the human to begin a process of complex change. We should consider the role and nature of additions and especially the idea of prosthesis. If we begin to analyse the phenomenon, we may be able to perceive better where such movements on a mass scale may appear. If we continue to accept as inevitable the provision and promotion of new devices until we have no option but to participate, then we will find ourselves very soon with implants. We move from plantations to implantations.

Prosthesis generally refers to putting something onto or into the body often as replacement. In particular, creation of artificial limbs comes to mind. But the meaning is wider, such as in language, and implications are broad. Common prosthetic devices could come in many forms, from wooden legs to dental implants, glasses, pacemakers, artificial eyes or brain implants in relation to human cyborgs. Some argue that humans have always used tools and that technological additions are the same. But technology that can change the user in direct and fundamental ways is qualitatively different. We could also consider plastic surgery as an increasingly common intervention into the body. Some surgical interventions are reconstructive and some are cosmetic. A sufficiently deep degree of prosthesis at some stage will transform humans into something else.

So in one sense humans have long utilised prosthesis. However, there is a distinction that might help us. In one category of prosthesis there is just an effort to restore, recuperate, replace, recover or regain something lost and rehabilitate the body. That might be termed classic prosthesis. In addition, there is the effort to extend normal, expected, ordinary capabilities of the human body through extraordinary, technical enhancement such that there is an improvement beyond normal expectations and organic evolution. Additional, extraordinary enhancement is the essence of transhumanism. Such augmentation represents deliberate interference in the natural condition and may even be calculated to end evolution in the biosphere. Intervention through addition in the manner of artificial selection is a serious involvement and represents a high degree of attempted 'enhancement.' Most enhancements initially will be individual and not necessarily heritable unless involving genetic engineering. Nevertheless, significant interference in the human mind may condition natural evolution. Eugenics returns.

Enhancement of individuals may direct evolution in a particular way. Certain traits will be enhanced, notably the one which desired to create the very conditions of enhancement. The materialist worldview will create more of the same. As there is an investment in one element of human development, there must be a number of costs. We will begin to have less accommodation of non-materialist perspectives of the old type. The sense of idolatry of quantification and implicit focus on measurability of things, substances and materials will cause a de-emphasis of qualities. Qualities that are outside accepted frames of references will be then compromised and diminished. Conditions of scientism and commitment to technological existence over existing human potential will prevent evolution in an organic way. The inherent weaknesses of the materialist mind-set that have contributed to capitalism and communism, commercialism and consumerism will be replicated in the surviving capacity of those forces combined and channeled by the tech-transmutation of the human into a machine. As hi-tech corporeal control increases, human qualities may decrease. There is a trade-off. Betting on the power of technology involves a failure to accept huge costs of technology to the biosphere and ethnosphere. The technosphere then will triumph over humanity. The ideology of scientism promotes a policy of transhumanism, not in the classic sense of restorative prosthesis, but rather to enhance the human technically so the amount of biological humanity is decreased and possibilities of predictability and manageability increased. With enhancement of computation and interoperability you will be habilitated or qualified for the technosphere. There is not a sliver of doubt that the possibility of the end of humanity is nigh. Nuclear war and bio-tech accidents represent the greatest threat. For some others it is environmental degradation. I suggest that some people would survive even in the worst scenario of cataclysm. But if we change our very nature we cannot survive as the same species. Certainly we could 'survive' as some machine, machine-hybrid, cyborg or as a nodal point on a digital, quantum, biological or quasi-biological network. So this is arguably the greatest threat to our species. We are told that this may be the last of humanity. We are told that we must just transform. At the same time we are told that technology will bring us to the promised land. These two propositions point to the carrot-and-stick approach for us dumbed down donkeys. Transhumanism involves mechanisation of the human body or hi-tech enhancement thereof. There must ultimately be a transformation of its nature. As articulated by Julian Huxley in his 1957 essay,

'Transhumanism' relates to a transcendence of human nature. It can be found in particular forms in famous documents such as the *Transhumanist Declaration* (1998). It sounds good and even balanced.

Posthumanism is a different or later concept whereby the construct of humanity disappears in some way. They may be related or entirely separate depending on the context. Certainly there could be catastrophe but that might also suit some adapted humans. As transhumanism is the pressing issue and more than likely leads to posthumanism, the idea of the former should be examined principally. Posthumanism purports to be a reaction at times to the supposed confinement of humanism to white males and people of power and exclusion of other beings and animals. In this form it may be as aspect of an ideological strategy which is materialist and shapes its arguments accordingly. It is curious how an ostensible base, for some, in enhanced recognition of animals seems to lend support to a process of bio-experimentation with hybridity. The implication really is that such philosophical policies are hand-maidens to materialism, science and scientism. For people like Katherine Hayles, it grew up as a critique of the rational, liberal humanist, autonomous subject out of the Enlightenment from philosophy, literary studies, science, robotics and artificial intelligence. A definition might be as follows. *Posthumanism refers to a range of philosophical, political and practical approaches that seek to move beyond humanism narrowly construed and past personhood, sovereignty and agency and thus allows, anticipates and promotes deconstruction of the human and their exclusivity for a variety of purposes that prefigure eradication thereof or of boundaries, totally or in the way they have existed, by treating or making individuals more embedded, allowing elevation of other beings and hybrids, integrating them as nodes into networks of information.*

Transhumanism is primarily related to transcendence of physical limitations by technological enhancement and purports to retain the humanist element. Huxley talked about transcending religion, law and social structure. Posthumanism is partly a deconstructive movement to promote and parallel what is going to happen in the hi-tech domain. The better view might be that technological enhancement creates changes that will inevitably alter our conception of humanity and thus represents a transitional stage to posthumanism prefigured by morphological modification. Transhumanism is related to posthumanism or there is a big logical connection. 'Trans' suggests transcendence primarily but also 'transitional.' The idea of transcendence suggests a material overcoming of physical limitations such as death, disease and disability. This is implicit in the work of people like Max More. His assumed name is suggestive of what the philosophy is. But the sense that this is merely a transitional phase is part of the true purpose. Once it starts it is unlikely to stop. If we look back at the influential futurist, J.D. Bernal, it is clear that humanity as we know it would be replaced and that there would be some interim period.

A sense of trepidation was also evident in the work of Theodore Roszak (1933-2011). His book on *The Cult of Information*: *A Neo Luddite Treatise on Hi-Tech, Artificial Intelligence and the True Art of Thinking* (1994) and the information-processing section criticised a cult of computation. Another earlier prescient thinker was

Günther Anders. The impact and process was identified by Anders in his essay on tv in 1956 called 'The World as Phantom and as Matrix.' He believed that the dispersing impact of tv in the home was turning us into a homeworker for the system that made us into 'mass man.'

> *"To complete the paradox, the homeworker, instead of receiving wages for his work, must pay for it by buying the means of production (the receiving sets and, in many countries, also the broadcasts) by the use of which he becomes transformed into mass man. In other words, he pays for selling himself: he must purchase the very unfreedom he himself helps to produce."*

Anders pointed out that there was no need for Hitler mass rallies. Technology allowed people be at home and be conditioned under the illusion it was fun, private and without impression of any sacrifice, according to Anders. The balance of power has shifted away from the citizen and we have ceded so much that our choice and control has been compromised forever. Once choice has been excised through relinquishment of control to people who have commanded it, we cannot expect a good outcome, especially when we have also yielded rights. When the roles have been reversed and it becomes apparent that our liberty was an illusion and that such freedoms as we possess are only those which the powerful give to us as their gift for obedience, then all their warped fantasies are available to them at their leisure and pleasure. But how can an apparently rational practice become a threat?

Network Transhumanism is the biggest threat. This to me is where converged and powerful networks, on which we are dependent, seek to integrate us into them through implants and interventions in our body to add to their algorithmic, reflexive impact on our mind. You are usually presented with a picture of humans walking on water and the blind being healed. Zoltan Istvan is an example of a proponent of transhumanism who uses this argument. However, it is more likely that the tech-concentrating elite will gain access to all the benefits available including a ready resource of organs for transplants and cells from other humans as they see fit. The ruling techparty can enjoy such modifications and power as all despotic elites achieve. The mass of humankind will merely feel a gradual or sudden takeover of the body and mind by technology. Firstly, it can occur through dependency on networks. The choice of opting-out becomes less and less. Then proximity to networks is increased. Pervasiveness of technology makes escape impossible. You must make yourself available to the networks. This proximity increases until you effectively will be joined with the technology through merger. Dependency-proximity and pervasiveness-merger allows the individual be changed by being linked into the network. Once an individual is linked into the network, they are minute scraps of a system of control by physical, computational power greater than them. They are increasingly assimilated into the technosphere.

Posthumanism being a diverse range of ideas deriving from certain thinkers like Pepperell, Hayles and Haraway may relate to consciousness and identity in a new relationship to material. Disaggregation of humans may allow distribution of consciousness or maintenance of distinctiveness in different form. It may be purportedly driven by concern for animals, nature, health, safety or other ethical claims. It is difficult thereby to conceive how humanity retains today's essential nature. It is related to postmodernism and deconstruction. Earlier thinkers were aware of this and McLuhan (1911-1980) indicated that we would be the sex organs of the machine. While posthumanism is said to be distinct from postmodernism, the impact is similar. Concepts of the human, individual and person are dismantled. Belief we were given the 'individual' concept by the Enlightenment arising from rational humanism also suggests it can be taken away. Personhood is not constructed but real, epistemologically and ontologically. We are, existing, and know that we are and know that we know that we are. Posthumanism may disassemble us and give the pretext for our actual dissembly and this dissimilation or making different will be accompanied by much dissimulation or deception. You will be told this is inevitable. While posthumanism claims to be motivated by problems posed by humanism and includes an ethical claim to repair, restore or alter, it has the same sense of entitlement to interfere in a profound way as transhumanism. It sees an end of a humanist metanarrative that is anthropocentric and is thus postmodernist. By a Procrustean strategy, if computers are described as logical or rational, that aspect of human nature can be regarded as the essential part and the claim that essential human nature persists is not apparently dishonest. A punt on posthumanist 'evolution' is a monstrous gamble for humanity rather than the real thing involving natural selection scientists used say was critical.

Posthumanism is possibly a political tactic to lubricate assimilation of the populace into the technosphere. That some advocating such theories do not understand the puppet-mastering of them for political or ideological purposes is possible. Some others are part of the great death-wish cult inherent in the disenchanting, nihilism unleashed by the Enlightenment and consonant with a certain magical stream in the Scientific Revolution that is partly inspired by a Faustian risk-taking spirit perversely promoted as liberating Prometheanism. Posthumanism appears to draw on a wide range of sources to create a kaleidoscopic view of the human so that the only constant is the phenomenon of change at the behest of controllers of perception. Like transhumanism, posthumanism will leave humans gasping for air. It represents in effect a philosophical and practical proposal to deconstruct humanity and we will suffer. Why this strange force persists is perhaps answered by the mesmerism of materialism and a false promise of the real power of technical tools. Scientists seem willing to risk our future on their emotions of pride and power. That this force exists is undeniable. It is not stated intentions we must examine but the actions. The front side of transhumanism is assistance for people who are lacking some physical base. Restoration is fantastic. The back side is the use of low-level technical 'enhancement' or implants to control the populace as subjects or tech-slaves in the Empire of Scientism.

Anti-humanism obviously drives the transhumanist and posthumanist force. The results of this movement are enslavement or demise of human nature and nature itself. Promissory science, promissory technology, promissory re-engineering of humanity will finally fail us and the environment. We must take and bear risk that others have chosen for us often based on their mere curiosity or misapprehension caused by materialist belief systems. Why we should allow others determine the future of the human race because it suits their disposition and ideology is beyond comprehension. We should not play dice with our future for the sake of certain philosophies or political ideologies that preach a proactive approach without prophylaxis. We are allowing ourselves be re-defined into obsolescence. This is a deliberate assault on the human race and not a mere concatenation of unavoidable, disparate factors. One consequence will be that we shall become lesser through the active de-construction and re-construction of who we are. This threat is made real because we have lazily ceded long-established notions of our humanness to ideologues and people with a vested interest.

Olaf Stapledon (1886-1950) was part of the scientific movement in the early twentieth century that anticipated evolution of humankind through cycles of catastrophe and then constructive or artificial evolution. Before many of the science fiction writers got in on the game, scientists such as Bernal and other philosophers and writers like Winwood Reade (1838-1876) anticipated and determined this direction. In doing so, the word 'evolution' must appear inappropriate insofar as it represents a much more deliberate process. Being deliberate it must employ the essence of such rational development. It is here that hubris comes in and over-reliance on a particular instrument of thinking becomes the exclusive direction. The mind that thinks solely mechanically as a tinkerer is prepared to intervene in complex patterns with some clear sense of certainty in their own mind. Where tinkerers get together to think, the tendency to unskilled experimentation deriving from real knowledge and a muddling-through mentality may produce catastrophe. How do scientific tinkerers get the wherewithal to decide on the future of our species without consensus or consent? It is not any natural evolution but takeover of the vehicle by a small group of people who tell us as they accelerate that it will be better for us after they drive over a cliff.

Transhumanism moving towards posthumanism may be associated with postmodernism and deconstruction and with thinkers like Derrida (1930-2004). He often speaks like a hypnotist. The approach of certain philosophers represents something akin to hypnosis in style. Deconstruction is associated with philosophy departments and certain, strong ideological trends but the greatest deconstructive force is scientism. It breaks down what was there in order to establish scientocracy. As Wells indicated, 'false loyalties' must be broken down. That meant that nothing but scientific corporatism, curiosity and control would remain. Even without this movement, technology leads to less humanity. Humanism was based on a religious ground. The Enlightenment notion was based on spiritual context. It did not require Adam Smith and David Hume directly be religious as their views operated in a society where such assumptions were embedded. Recent books such as *Religion and the Rise of Capitalism* (2021) by Benjamin M. Friedman indicate and explore this important connection. Many schools now reduce history to correspond with their ideological agenda and this happens to correspond very well with scientism.

Pharmaceuticals are also prosthetic. Terence McKenna (1946-2000) likened psychedelics to prostheses. There is something in that. Pharmaceuticals make the body able to do things it could not. Some interventions are restorative but many may alter the functioning through enhancement. If we think about The Pill, it not only facilitated birth control but is associated with the hippie movement and radical social change. It undeniably adds something that then interferes to stop what was hitherto normal. So interventions that are sold as enhancements may operate to prevent or immobilise some function in the body. Thus enhancement may involve impairment of function. All such interventions have costs. While classic prosthetics, in the sense of replacement and restoration, have been around for thousands of years as artificial toes attest, the degree of transformation now is formidable. In addition, if we take the pharmaceutical example, it is clear that programmes and policies have been very cavalier with public health and have sacrificed people for profits. The profit-making force is a powerful motivation to abuse. Pharmaceuticals and prostheses of the nature of biomedical implants invite rent-seeking. Mass products make massive profits at our expense. Whether it may be psychedelics, pharmaceuticals or prosthetic implants on a nanoscale, we will be told of the advantage, utility or necessity of their wide-scale use. Do not expect such forces to have interests of the person, as top motivation. Certainly advantages and benefits are usually exaggerated and costs often ignored. Pharmaceutical posthumanism through nanotechnology and smart substances represents one of the greatest threats to human freedom and existence.

Accepting that technology intervenes in the biosphere with huge costs and consequences and in the human body, mind and soul, we must be aware of its potential to de-humanise and de-personalise us. While medicine may restore us, enhancement will add something beyond what was normal or within reasonable expectations at the time. Enhancement has costs as well as benefits. Enhancement may be in solid, liquid or some other even gaseous or plasmic form. Enhancement may impair and take away function deliberately or accidentally. The purpose and intent of enhancement is critical. If enhancement is pushed via a mass, top-down, compulsory scheme, then the danger of an impairment of function increases. That which impedes the person may be seen to improve some other objective our puppet-masters have promoted. Advocates of transhumanism, outside of the weakened and disabled, are usually powerful. Enlisting the unfortunate and impaired should not blind us to the reality of forces promoting the general agenda. The propaganda machine knows well how to programme us. In a world where information is increasingly regulated in favour of our invisible governors and wherein power of instruments of subjugation grow exponentially, do not expect some cherry-picked promises of the propaganda machine and associated media mesmerism, with an aura of authority, will permit you apprehend the practical truth. We should evolve spiritually anyway. Costs of failure will get higher as we rapidly become hostages to an ineluctable agenda. Technocrats and professional thinkers decide to tinker with our body, mind and spirit. We are at a crossroads if not already on one path. Crucifixion of our species incarnate is not of a type that involves a resurrection.

The greatest prosthetic on the body politic is the system of government or governance through corporation. When it becomes more than it should, it assumes more power than is reasonable and proportionate. Then the body of any corporate entity gains a life of its own. When it is an oligarchical power centre, persistence and perpetuation thereof becomes more significant than the people whose good the structure pretends to satisfy. The argument that morality itself should be transcended for some purpose like oligarchical government or the governor's interests can be seen in Nietzsche, Machiavelli (1469-1527) and back to Callicles in ancient Greece in the dialogues of Plato and Thrasymachus circa 400 years B.C. An idea of 'might being right' is very worrisome even when it is packaged in persuasive, hypnotic policies. It is more concerning still if predicated on a sense of superiority. It is more worrying yet if superiority embodied in a mode of governance can utilise technique derived therefrom to project sophisticated propaganda which reframes what and who we are. Transhumanism may entrench oligarchy with cherry pickings for the top and chains for the bottom. It is no surprise that the monster of governance wants to control by magnetising and managing.

That which is added on to us may not be an act of mere restoration but rather fundamental alteration of our nature masquerading as enhancement. The augmentation is driven by the same force which breaks down, deconstructs and destroys. It will banish the essence through reduction and claim it never existed thus proving the presumption which prompted it. The tools and symbols that create an illusion of neutrality and objectivity, conviction of a sense of elect entitlement, obsession with predictability and power provided by mathematical and scientific formula of undoubted efficacy are then paradoxically applied to sacrifice the entity which created them. The by-product is confused with the producer. Toys and things can be tinkered with to produce the illusion of exploration to replace and reduce the human to a marionette. That which may appear enlightened may present a dimmer possibility or effect. Thus transhumanism is not a stand-alone movement of the nature sold to us. While it may have been energised by the push for immortality for people like Ettinger, the prior basis of movement can be located in a drive for scientific domination as explained by Bernal and others. It is also possibly part of the force which can be traced back through Nietzsche to ancient Greece. It has the imprint of the magician and desire to conjure new beings and destroy, imprison or transform existing ones. Even respectable, well-researched, standard scientific thinking may cause drastic change. One example of the latter is the implications of conscious realism or agency. This will subtract to allow additions to you.

Thus the human in a technological environment is 'enhanced' or mechanised by prosthesis. The argument is that it is an example of individual freedom to change human nature and shape even though it may alter our type of being. The possibility or probability is that governance will utilise technology to incorporate humans into the network leading to a posthuman situation. Whether it is associated with a desire to control other people or a function of the nature of technology and technique, humanity will be assaulted. Scientists and philosophers are telling us humanity is nearly over. The promise and playfulness suggested by transhumanism obscures power of the ringmasters using scientism as an instrument of control to create a fundamental transformation. Such hi-tech transformation of humanity, consistent with machine misanthropy and an obsessive cult of management, metrics, information and elite experts that operates in transnational spaces through co-ordinated networks and technical intervention, is on the way. In all this, there has been a failure to comprehend or accept the nature of the human as a whole person that is fundamentally part of the natural environment. The greatest impact of philosophical transhumanism will be in promoting technique to allow humans be transformed to adapt for the technology they will become. Such deliberate tinkering will be called evolution, despite being a conscious attack on natural selection. Freud said that man was becoming a 'prosthetic God' but was unhappy with it. Prosthesis is not for mass benefit but a Trojan horse. Scientism is akin to magic. Renaissance and Enlightenment humanism sought to make humans godlike to replace the divine. It seems to have sought to bring god back as a created, mechanistic form of perfected computation, ratiocination to control free will. While there are idealists who hope for positive things from 'existential posthumanism' without hi-tech reliance, that may not happen. This evolution is inversion.

The notion of prosthesis is misused for transhumanist propaganda. It is argued that common additions to the body represent transhumanism, if they merely seek to restore some function which was lost. However, this is not strictly an extraordinary enhancement over such functions as are within normal or ordinary expectations for humans. Impetus to restore function did not come from any theory about fundamentally altering the nature of humanity.

There is an attempt to label everything in our environment that we use as tools as prosthetic. Thus Freud is cited to the effect that once we wear clothes or use tools, we are engaging with prosthetic devices. This fails to distinguish between material which is permanently integrated into our body and mind which we adapt to and external instruments. In this vein, there have been recent attempts to label everything such as clothes or literature as prosthetic. Certainly books like *Moby Dick* (1851) explore prosthesis but it is taking it too far to extrapolate to some general theory of prosthesis therefrom as some do. Such extrapolation seems more like a process of normalisation of theory to lubricate more specific ideas of prosthesis that are clearly artificial or extraordinary and obviously intended to alter the body in some lasting fashion and probably on a mass scale. While writing itself is a significant technology that has altered our culture and mind, it is not the same as a technical implantation in the body that augments function through mechanical means. There are debates about whether prosthesis and simulacra or imitations are the same or not. It is clear that prosthesis can make likenesses that become original but not the original. Hayles believes that we have already become posthuman. The body itself is supposedly prosthesis. Some posthumanists link to postcolonialist discourse.

There are many who even claim that posthumanism is merely a reaction to transhumanism as well as humanism. While posthumanism clearly attacks anthropocentrism and the centrality of humans and humanism, it is less persuasive as a critique of transhumanism. That seems to put the cart before the horse and may represent a misapprehension of the force of technology and technique.

The idea that we may improve the world in some way by deconstructing positive conditions, potentialities and possibilities for humans that are calculated primarily to protect them against misuse of arbitrary power by yielding more of our sovereignty to omnivorous concentration and orchestration of techno-governance seems to be based on a mistaken analysis, foolish or else a front for the would-be governors promoting confusion in the guise of enlightened elevation of consciousness. Some thinkers assert that humanism has become identified with pure rationality. This is true insofar as the Enlightenment eventually excised esoteric and Divine presuppositions of the concept and logos was lost. It is a serious strategic and actual intellectual error however to compound such a loss by shifting to systems which want to throw the baby out with the bathwater. Reason destroyed broad humanism and gave us the destructive materialist mentality. Now we are to deconstruct the pale imitation that persists, to go further from the lighthouse. We are either to discard rationality as some posthumanists plead or to privilege it as some transhumanists propose. Both make humans suffer for the promise of good treatment for other beings some not even imagined and many to be engineered by humans as the new God. I think posthumanism will not work for most perennialists who should reject engineered existence.

Thus transhumanism operates through extraordinary enhancement beyond mere restoration or treatment. Enhancement is through technological or mechanical means but does not have to be so narrowly construed. Chemical means may also enhance function. Likewise, addition may interfere with some function or provide some new ones. When everything can be construed as prosthesis, from language to art, then the true meaning of transhumanism is obscured in many descriptions of exteriorisation. It is not this inflated idea of prosthetics that should be considered in relation to direct alteration by the augmentation of *homo sapiens*. Contemporary and prospective transhumanism, based on a narrow definition of prosthesis calculated to enhance function beyond existing or normal expectations, represents a phenomenon of devastating, potential consequence. The worst impact will be relinquishment of humanity to a new reduced, ratiocinated, controlled and imprisoned consciousness that is a false substitute or simulacrum. Blake (1752-1827) realised this. Arguments that suggest everything has been like this before are false extrapolations whose effect is to lubricate the process of prosthesis and therefore to project, consciously or not, propaganda that seeks to persuade that transhumanism is nothing new. While posthumanism may have some noble aspirations, the most probable impact will be to form direct alliance with transhumanism, rather than fashion an antithesis. If one examines the origin of both, the most parsimonious explanation is deadly desire for control espoused by certain scientists. The chartered, measured and managed self is what we are becoming. The 'quantified self' will be measured in a digital mirror instead of being reflected in a karmic one. How the individual or group is presented or perceived will merely depend on the degree of distortion by rotation produced by projection of policy co-ordinates onto the object of control. We have yielded our future to a circularly sawn diminution of humanity by forces of rationality. We have forgotten that the person and their qualities are nor abstractions but realities of the essence of creative force.

Our lives are determined by obsessions of our own and others. The worst and best come from obsessions. Thus we must be careful what we and others pay attention to and pursue as policy. Thinkers like Gabriel Marcel (1889-1973) promoted personalism, understanding processes of alienation by technology and abstraction and why we need to seek communion with others based on vision, values and love. The propaganda model and machine will increasingly serve you add-ons by ads and inversion. You will be unfavourably compared with other cyborgs, androids, robots and the upgraded. That sold as a big plus to the individual may become a mass minus. That added-on will transform the human into something else. Science and technology are the governors. Ideas or pseudo-philosophies that tech-fantasies spawn are offspring of science and part of the ideology of scientism that informs the new, global imperium and power of the hi-tech enhancement emporium. Transhumanism will be prosthesis for elites but pacification, prison and permanent devaluation for everyone else, founded on a sadly reduced view of humanism and an indefensible idea of the human. Technophiles seek to create abstract solutions that may interfere inappropriately in complex circumstances. Such interventions are calculated to affect us that allow unchecked malignant growth of ratiocination and increased commercial concentration. The use of theories and fictions to represent the world includes the possibilities that they can misrepresent it. The study of semantics and the warning that that map is not the territory is a critical idea for scientific study. Limitations of the map are further limited in the hands of people removed from robust systems of reasoning.

Corporations have legal personality and their rights are rising as ours decline. A corporate entity can act as prosthesis also to personal will. Some theorists see the link between law and magic. Powerful legal personality may impose mechanisms of control by ostensible enhancement. Legal personality shifts us from spiritual transcendence to commercial transaction. Many people who studied law noted connections between it and magic through the power of words to achieve results, ritual, ceremony, performance and judgment. Legal personality will increasingly apply to animals and androids just as it had already done to inanimate objects and legal fictions.

Legal systems such as the Roman or Common law also regulated destructive magic. As Hegel (1770-1831) noted, magic has always existed. We can go back to Plato and beyond to read about its role. Issues that interest the transhumanists such as immortality, superhuman powers, superintelligence, transformative substances are the subject matter of much magic. In both there is also a concern about control of others. Note that the word 'intelligence' is also used in relation to military and security information about people. Thus superintelligence may refer to the degree of information held by corporations about us. The link between magic and the secret services is well-established. Magic was often seen to be about the gods or control of spirits and thus overlapped with religion. Similarly concern about misuse of magic always existed. While transhumanism may bear a striking resemblance to magic in its desire to transform, it may also share the disposition to destructiveness seen to be the dark side of magic from time immemorial.

THE SECOND PART

The Magic Spells of Transhumanism

It is as if we are being mesmerised by the spell and spectre of technology and scientism so that magicians manipulate us to manufacture a new machine to merge into a network. Some movements are orchestrated and some are just inevitable results of technique and the inherent nature of technology. This section examines how the human individual person may be reduced and then deconstructed by scientism and hi-tech networks. The idea of conscious agents, for example, may prefigure radical re-alignment of our assimilation into the technosphere. This threat requires us to re-examine or re-affirm the nature of the human individual person. Ideas such as 'personalism,' 'holism,' and 'phenomenology,' which gave rise to concepts of the whole person and significance of the entire organism, oppose disintegration of the idea of the human. The phenomenon or motif of magic, by analogy or actually, is part of the possible explanation of this field of contest. To some extent, technology and transhumanism is promoted by illusions or tricks on a public largely entranced or mesmerised by media and manufacturers of the consensus matrix. The spell is cast through inversion, incantation and mantra which accentuate benefits creating fear of natural limitations while reducing conscious agency to mere computational power and ratiocination.

Reducing Persons to One of Many Conscious Agents

"As Gregor Samsa awoke one morning from uneasy dreams he found himself transformed in his bed into a gigantic insect."
Franz Kafka
The Metamorphosis (1915)

The frequency of use of the word 'transition' in the transgender context demonstrates a graph consistent with use of the word 'transhumanism.' They may move closely and in parallel as the work of Martine Rothblatt indicates. I make no points about transgenderism at all. However, there has been a process of acculturation in the world to the notion of transitioning as a matter of identity. Without probing the connections in this correlation, it is clear that we have been much more accustomed to the notion that the condition of birth or a result of later development is now regarded as more malleable, consonant with wishes of the individual and technological capacity to effectuate physical changes to match a perceived state. The deeper sense of general morphological plasticity engenders a disposition that will be more predisposed to technological intervention to achieve non-gendered transformations. Making a choice from deepest motivations of one's own free will can be distinguished from an imposed change by compulsion, calculated to achieve some general objective supposedly based on the common good, latest scientific modelling or inflated threat, whether it be large or ostensibly very small.

There is a sense of a phase of humanity in transition. This is appearing in philosophy and practice. It represents a move away from humanism. Humanism, as indicated, is seen to be a specific notion that originated in the Renaissance. However, individuals realise that they are humans. It is not a mere construction, invention or mental trick or aberration. It was not dependent on the writings of philosophers. We know we exist as humans and persons with consciousness. We should resist the scientific and philosophical process of deconstruction, de-humanisation and de-personalisation. As persons, we experience the phenomena and qualities of consciousness, personhood and selfhood. Linked to this intellectual attack on our very existence and nature there is persistent undermining of perception of reality. The impact of such supposedly scientific and authoritative claims is to create confusion. Confusion is a strategy of certain modes of propaganda, programming and the process of normalisation to new norms. One strategy of Enlightenment science, as revealed in the 19th century, is to deny some phenomena associated with humanity. By denying them, scientists shift the burden of proof onto people who possess, argue for, admit or promote the actual phenomena. Thus scientists deleted spirit from discourse and denied spiritual phenomena and then required proponents to prove it whilst even telling them that they would be uninterested in proof of spirit or spiritual experiences even if it was presented. Now it is worse. Having taken spirit out, some want to delete humanity, free will, individuality, personhood, selfhood and consciousness. If the average person accommodates this they become guilty of acquiescence unfortunately consistent with acceptance of their inevitable demise.

Again, there are a number of notions of humanism. In its simplest form it refers to a focus on humanity. More specifically, it indicates a tradition of thinking about people that emerged from the Greco-Roman, Judeo-Christian and indigenous traditions. The extent to which it grew up during the Enlightenment has been exaggerated. On the contrary, the Enlightenment began a process of disenchantment and facilitated a process of reductionism that caused disillusionment and nihilism. Pico della Mirandola (1463-1494) is an example of someone who had a whole picture of the individual. The idea of the human as a reflection of the divine force later became replaced with secular ideas. Notions of liberty, equality and fraternity became identified as non-religious ideas beyond the divine source. Inalienable rights however were justified by higher sources in the US constitution and Deism. This was based on reason. In the twentieth century and this one, the assault of science on religion has had an effect of implicitly undermining the basis of rights. The assumption and argument is that reason can answer all issues and objectives can manage us. Tool becomes theory. The source of such objectives is assumed to be implicit from the perceived direction or interests of science. Hence a combination of ideas of evolution, curiosity and desirability of technical intervention to solve problems or merely because there is an opportunity, overrides other approaches.

Science has attacked the spiritual composition of humanity very effectively. In the late nineteenth century in London, Thomas H. Huxley and others eradicated the spirit from much scientific discourse. In such contexts other substitutes were 'forces' instead. Thus psychic forces and telepathy were terms coined because of the unwillingness to recognise what hitherto were spiritual. Psychology and philosophy have continued this scientific deconstruction. Neuroscience makes it worse now. In this technique of mind-control, the human is reduced and disassembled. Far from being a spiritual creature of divine nature, the human is increasingly regarded as a mere construction, like a delusion, possessing no free will, nor self and certainly no soul nor spirit. The human becomes a burden on the planet to be managed by technology.

The 'conscious agents' idea is interesting but the implications are worrisome. If we reduce humanity to conscious agents then we are merely lower life-forms in a hierarchy of computational power. Changing or reducing definitions to how something behaves diminishes us. Donald D. Hoffman wrote *The Case Against Reality: How Evolution Hid the Truth From our Eyes* (2019). His argument is a sophisticated and interesting one. He does not use the term *qualia* but 'conscious experiences.' Hoffman uses a metaphor of computer icons to indicate how we perceive the world. His ideas were worked out *inter alia* through dialogue with people like Francis Crick. Crick famously argued *"You are nothing but a pack of neurons."* Hoffman relies on the work of Bogen and Vogel and how the brain works when surgically split in epileptic patients. He examines evolutionary theory and lessons from studies of visual perception. He uses studies to indicate that natural selection does not shape us for veridical perception and explains how we perceive based on fitness before truth.

As well as using a recurrent computer metaphor of icons in argument, Hoffman relies on information theory which underlies communications technology and information studies. Ideas such as computation and data compression are applied to visual perception as are ideas of 'hacking,' visual code, screens, networks and heuristics and evidence from computer simulations. Hoffman speculates on the possible analogy of conscious agents interacting and cryptocurrency. He gives examples of the utility of such information in the marketing context which he has employed in helping a major company for example. He argues for conscious realism and that our 'interface' does not point to objective reality. We are conscious agents, one of potentially many. He defends himself against the argument that this divorces him from living consciousness by emphasising 'fascination with the living subject.' A conscious agent is dynamic and it perceives, decides and acts. This allows him use the idea of measurable space and a Markovian kernel. The definition of 'conscious agents' is 'just math.' There are some good parts which accord with the mystical, spiritual or perennial tradition, although he probably rejects this. In relation to cognitive neuroscience he says at page 190,

> *"This framework does not assume that biological neurons and their networks are the building blocks of cognition. Instead it takes consciousness as fundamental and then has the task of showing how spacetime, matter, and neurobiology can emerge as components of the perceptual interface of conscious agents."*

He indicates, at page 196, that the novelty of this theory is to draw from old ideas of philosophy and religion and make them testable and,

> *"allows the ideas to be refined under the watchful eye of the scientific method."*

He accepts that science is not a theory of reality but a method. He argues that if science cannot describe who we are, imprecise language cannot either. He agrees with Dawkins. He goes on at page 200 to suggest that we can,

> *"foster... a scientific theology, in which mathematically precise theories of God can be evolved, sharpened and tested with scientific experiments."*

If I was to identify one particularly relevant assertion from his work, it would be this at page 190,

> *"Conscious agents can combine to form new conscious agents, and these new agents can again combine to form yet higher agents, ad infinitum. When two or more agents interact, each retains its individual agency, but together they also instantiate a new agent. The more each of the agents in an interaction can predict its experiences from its actions, the more integrated is their joint dynamics and the more cohesive is the new agent that they instantiate. The decisions and actions of a higher-level agent can, in turn, influence the dynamics of the agents in its instantiation."*

More importantly, it is clear to me that conscious agency or conscious realism theory represents an ideal framework for a transhumanist and posthumanist society. Hoffman makes clear that while consciousness is fundamental, human consciousness is *not* fundamental or distinctive. He rejects the idea that this is magical or Promethean. However, it is clear to me that 'scientific theology' could seek to continue from where dispiriting finished. It can shift the burden of proof until all that might be left is the new god of AI networks in which humans must interact, with no special status, no distinctiveness and huge disadvantages. While Hoffman rejects the argument that the reduction of consciousness to mathematics is not a denial of first-person consciousness, he seems to do so on the basis that scientists are engaging in a first-person way. When we look at the first person consciousness of other scientists we should be worried. Hoffman is a courageous and innovative thinker. He is linked to Crick. Crick called Bernal his 'scientific father.' Bernal envisaged that humans who would not change in the future could subsist in a 'human zoo.' He too had a healthy interest in humans as subjects for experiments. While this may be 'post-materialist' science, it may also be posthuman. Time will tell. It could also represent a scientific expansion towards spirituality and religion that is a takeover-bid as I have anticipated elsewhere. I do not accuse Hoffman of this nor say that this is his intention, but look at the likely practical effects from a pragmatic perspective. What is left out is subjective consciousness of the other and sense of significance thereof. That human consciousness is distinctive and is part of a fundamental consciousness is not the same as saying that consciousness is fundamental in another sense. All the time we are being forced to prove our human dignity and justify ourselves in narrow symbolic terms that must necessarily be subject to incompleteness while promoting elevation of machine-intelligence and AI as inevitably superior in terms of a narrow assessment or analysis thereof in computational terms. Hoffman's work, in my view, may represent a reductive confinement of human potential as a result of an unduly, narrow mathematical approach that confuses the proof with the phenomenon and applies the lessons of computer networks reflexively and recursively.

The mind-set that seems to drive the agenda of transhumanism and thrive in its rapidly-emerging milieu is one that is risk-taking while forcing individuals to be risk-adverse. It goes very far beyond Ayn Rand's (1905-1982) egoist, selfish entrepreneurship. Unfortunately, many problems the world faces are because of such a Monte-Carlo last-chance approach. While innovation requires risk, the idea of costs of production as externalities does not bode well. Benefits are constantly paraded while costs are ignored and risks downplayed in the promissory notes. We will certainly face many problems which were predictable but permitted. The threat of nanotechnology released in the environment combining with poisoned fruit of some lab somewhere that is squeezing some strange bio-agents from some unfortunate screeching creature we would have done well to leave alone, comes from the cavalier attitude of science. Not only do scientists and technocrats have a likelihood that they can produce problems, but science may have an incentive. Advocates of the technosphere, mechanisation and technologisation of human biology have an incentive to allow the biosphere be compromised. Scientists have not only revealed a lack of concern for ethics and distaste for moral principles on many occasions but they have also shown a desire to control and boss humanity.

We will be re-defined as something else as we are de-humanised. A very possible candidate is this idea of 'conscious agents.' A possible result would be that society recognises conscious agents of which humans are but one. Then we fall down a scale of significance and legal rights can be accorded to mechanical agents at our expense. With greater investment in automatic hi-tech robots and tech-enhancements of the human promoted by the patent system, more non-human conscious agents will come into being. We can expect an endless stream of hardware that proliferates in the technosphere as we retreat into our homes until we are mere nodes in some network. Machines work on objectives and there will be no influence for ordinary people on the governance systems of globetechgov. Deconstruction of legal rights renders the individual bereft of challenge or appeal.

One solution which would be classified as 'bio-conservative' is to maintain as much of what was there as possible without undue and speculative 'proactionary' risk-taking. In that sense, the norm would be the ante-human. If there is a post there must be an ante. Ante-humans may prevent anti-humans. Personalism can oppose de-personalisation. Personalism in different forms represents a focus on personhood and personality as a beginning and builds on consciousness outwards allowing phenomenology and existentialism and other forces to elaborate through empathy and recognition. Personalism is against the idea of integration of the individual into the collective which is the objective of extreme left and right materialism. Personalism may then qualify unlimited individualism. Kierkegaard (1813-1855) and Wojtyla (1920-2005) propose elements of personalism. It is often suggested that personalism runs between opposing extremes of totalitarianism and pure individualism. Personalism is manifest in the work of Martin Luther King (MLK) and represents a fundamental aspect of his philosophy and theology. Personalism posits a special idea of the human with unique dignity. By denying unique aspects of subjectivity and interiority or spirit, the application or projection of a taxonomy or classification of conscious agents could create a de-personalising, de-humanising result. If one has no conception of the person and our human dignity, there is no need to acknowledge, grant or vindicate rights. By tinkering with the notion of a human person or individual and seeking to deconstruct the idea thereof, science and technology as misapplied through scientism may render superfluous the uniqueness of humanity and condemn it to its doom. The individual person has sovereignty and free will that gives rise to creativity and capacity for meaningful relations with others.

A long litany of human failings, tyrannies, destruction, genocide and environmental damage often occurs through collectivism or extreme individualism as opposed to personalism. That we have failed to recognise human dignity on many occasions has also meant that we have failed to recognise other cultures that we can learn from. Persecution results from de-personalising other groups. The solution is not to double down on collectivism by using technology at the expense of the person but to reinvigorate the idea of the person. That involves recovery of the lessons of perennial wisdom and human spirit and a certain reactivation of the spiritual realm or animasphere. Underlying posthumanism and transhumanism is a great failure to accept ourselves. Acceptance of oneself and one's essential dignity, uniqueness and subjectivity that is capable of exercise as a creative and compassionate force, through its inherent ability to recognise the other's personhood, is the essence of genuine community. Any collective, de-personalised, mathematical, reductive, technological sense of consciousness is merely an elaboration of some sophisticated simulacrum that seems special in the eyes of those who value machines over humanity and see the individual as a flawed machine, merely waiting to 'transcend' its limited biological cage to enter entranced and beguiled into a new one, gilded, guided and guilded by the enlightened materialist.

The conceit of those who may be mathematically or technically gifted and have benefitted from their promotion of transhumanism is breath-taking. As often happens with the technical mind, it assumes that success in one context is guaranteed in another. The facility with which mathematics and other theory is wielded, by people like Bostrom, to advance theories of computer simulation may be consistent with the Procrustean tendency in science. If facts do not accord with reality they will re-configure reality. While rejecting facts beyond proof they may establish presumptions clearly in favour of equally unprovable and less probable hypotheses merely through the aura of the method employed. Elevating the authority of science, mathematics, quantification and technical solutions has allowed preposterous propositions attain credibility to the credulous like the emperor's clothes. The Quant Emperor of the Empire of Scientism has many gowns but some show their nakedness.

Many thinkers have expressed ideas about our need to evolve. Sri Aurobindo looked very closely at spiritual evolution. Most spiritual thinkers or leaders are talking about evolution on a spiritual level, firstly individually and then on a community level. Religion makes a mistake that mystery religions did not. The latter understood that experience must be individual to work. There must be an initiation. That initiation is not merely symbolic but involves an actual personal conversion, realisation or profound insight. Religion has crystallised often into an empty ritualisation. The adherent gets a sandwich with nothing inside. They are told of other people's experience. Methods or significance of direct, evidential experience are substituted with ritual that is managed for the benefit of a machine. Mystics understand the need for experience of the spirit or the Divine. The mystic has participated in a direct experience and not merely in an experience of someone else. Without the living spirit or light, religion becomes desiccated ritual, rite and relations of power.

Mystical experience defies much traditional scientific, empirical or materialist paradigms. It often simply falls outside them. Phenomenological method partly based on personalism however permits recognition of individual experience as valid and acknowledges the reality of it. This is similar to the pragmatic approach of William James (1842-1910) and the idea of 'radical empiricism.' Similarly, Kurt Goldstein who developed the idea of self-actualisation underscored the crucial role of personal imagination in a German book sometimes translated to

The Organism (1934) in English. When conscious agents are identified with mathematical approaches, they will miss ineffable elements beyond the comprehension of merely complex computation. Those elements missed may make the difference between consciousness as understood by scientists and the real phenomenon itself. Thus what secret scientists believe they will have cracked will be an impoverished version of full potential consciousness. Spiritual evolution recognises possibilities of higher consciousness that will avoid this Procrustean proclivity.

Furthermore, consciousness as calculation fails to see completeness of human potential. Nikolai Berdyaev (1874-1948) pointed to isolation from the spiritual by humanism commencing with Leonardo da Vinci (1452-1519). He wrote of the unquantifiable mystery of the Divine within us. Failure to see those higher forces contained in itself and its sources means the individual is misled by illusions. There must be some submission to the highest force. Humanism would fail and seek to become sorcery. This direction in humanism seeks to deny the Divine and the God aspect of humankind. Berdyaev realised that slavery could emerge from freedom in his books. He discussed this relationship in *Philosophy of Freedom* (1911) and *Freedom and the Spirit* (1927). Disintegration of the spirit was the real danger. He believed that divinisation by man could not be the same as divinisation by the Divine or the light that must shine within.

Scientists and mathematicians come to a conclusion that something is true when they can demonstrate the proof. They confuse proof with the phenomenon, model with the mode. They claim that what we experience is illusory and can demonstrate that with mathematics. In doing so they ignore a long-established realisation in personalism and phenomenology that such proofs are themselves illusory. The idea that mathematics has an independent existence is a Narcissistic flaw compounded by uncertainty. When Hoffman argues that we perceive the world in icons, he then claims that replacement icons he uses are somehow better and perhaps different. What's good for the goose is good for the gander. Personalists and mystics understand that there is a person within the subject, a seer that is beyond the purely rational. The rational and reasonable of the head is a crust or veil. Ideas of personalism also sought holistic approaches which denied utility of disintegrated parts.

Deconstruction of the individual person by science is ongoing. Neuroscience and some philosophers of science seek to do what philosophers have done. Shifting the burden of proof means you have to prove personhood. This astonishing state requires us to assert what a person and what a thing is. Such a stunning stupor have we stumbled into that scientists can demean personhood, selfhood or consciousness itself. William Stern (1871-1938) distinguished between persons and things. A person is a whole being greater than attributes or the sum of its parts. Being such a whole person they can be self-directed in a way that other things cannot. William Stern was the father of Günther, whose surname later became Anders. It seems to me that the answer to many of these questions come from the basic concept of what a person is. Furthermore, the person inter-relates with the biosphere. They are organisms that work in that way. The tendency towards isolation and alienation from the environment is part of that viral de-personalising, de-humanising trend. Technique and technology tend to the technosphere and a totalising force that will be totalitarianism. There the person must become a machine, integrated by degrees. But once the first step is taken there is a great shift that will become irrevocable. The person will become a thing. Critical personalism of William Stern steered between the personalism which separated body and mind on the one hand and an attitude which broke down the person into a collection of attributes. We are not mere conscious agents but are greater than that. Even if one does not accept that you are of divine consciousness, the nature of personhood is distinct from lesser arguments or techniques produced by them that are things even if potentially described as 'conscious.' We must be whole persons again to combat post-personhood.

All serious theories and technological approaches should be tested in the light of some fundamental issues. What impact does such an approach have on the biosphere or organic life? Does this increase the technosphere or make us more dependent on technology for our existence? Does this idea serve humanity or threaten to enslave it? Does any such idea lead to a reduction in the dignity of the person? The answer to such questions does not preclude or impede scientific discourse. It is rather to have a better debate at the formative stage in order to appreciate the full implications. Transhumanism approaches are new in one way but may be representative of the totalising tendency towards techno-totalitarianism that others have warned about. That does not mean that individuals promoting such ideas or theories as rational explanations are responsible for all implications but the pragmatic approach asks about likely consequences. These questions relate to ideas of who we are and what emphasis we give to our body, mind or spirit. Even if science wants to dispirit us, spirit will assert itself until trampled out. In contrast to Hoffman, Federico Faggin also has a theory of human consciousness but includes identity and meaning. Otherwise conscious agency is a pale attempt to dis-spirit.

All ideological re-definition of essential human identity may be used to de-humanise and de-personalise. De-humanisation re-defines identity. Re-categorisation allows different policy. Resist denial of your humanity or you will pay a huge price for the alleged prize of enhancement. You may give up qualities. That which appears to be more may reduce relative importance of something else, undervalued precisely because it is not quantifiable. You risk accepting a Trojan horse to remotely control you directly or through dependency. Dependency could be part of the ubiquitous, predatory instinct of humans or a tyrannous plan to exert dominance. By reducing humans to mathematical constructs, scientists can dismiss elements consistent with higher consciousness while claiming they are exploring them. Reducing rationality by ratiocination through subtracting means scientists can propose more as inevitable enhancement while ignoring other inconsistent, unquantifiable qualities until the claim thereto becomes impossible. The ghost in the machine will be replaced by a *deus ex machina*. The god of science will come from machines like a theatrical device. It too will be a networked machine and we will be ghosts. The destructive magical process of inversion permeates the tactics and policy of science itself when it permits confusion of reflection with reality.

Spirit-Mind-Body and Boundaries

"And now the forces marshaled around the concept of the group have declared a war of extermination on that preciousness, the mind of man."

John Steinbeck
East of Eden (1952)

The de-personalising, de-humanising assault on the human, individual person through reduction, denial and deletion is permitted by attack on a proper concept of reality through de-construction. We live in a world of spectacle and sentiment which cultivates a culture of victimhood to prepare the subjects for a sacrifice. The original meaning of a victim is someone prepared for sacrifice. Curiously, this is part of the contemporary cult of victimhood. In a world that focuses on spectacles and charades of futile fun through bread and circuses, a parade of the defeated is part of an inculcation of people who present no opposition to the propaganda model that perpetuates materialism. The potential antidote of the animasphere is dependent on a full concept of the whole person. Recognising the whole organism of the person without reduction as a pluripotent being of consciousness affirmed by a persistent phenomenon of the perennial philosophy, wisdom or spirituality, naturally embodied or incarnated in an environment, that Darwin told us we were adapted to, is a necessary position to avoid the attack by a utilitarian, ratiocinated mind-set weaponised against its own source as part of self-awareness

If you assumed that spiritual people would be against transhumanism, you would be wrong. 'Techgnosis' comes to mind. In Japan, robots may be seen as spiritual by some. The relationship is complex and forces one to engage in ideas associated with the spirit or soul, consciousness, mind and body and relationship between them. My general preference is to use the word spirit or spiritual consciousness to include the soul, though I am aware of the distinctions. Where people do not use spirit they may accept psyche or consciousness that is similar in many ways without religious connotations. Consciousness had become an arena of contest in science and while advances have been made, they are still far short of any full explanation. Spirit and soul have largely been excised by science from their discourse in favour of psyche and mind. But there are immediate fundamental issues. The nature and existence of the mind is assumed by us today despite the fact that it was arguably a construction of the Enlightenment according to George Makari in *Soul Machine: The Invention of the Modern Mind* (2017). Also for example, an important proponent of transhumanism Kurzweil has written *The Age of Spiritual Machines: When Computers Exceed Human Intelligence* (1999).

Within this debate is a conflict between the idea of the body as a machine and brain as a specialised machine or computer. Identification of phenomena often depends on one's starting points. If one only trusts measurement then one will rely on things that are measurable and disregard the immeasurable and qualitative. This also involves the debate about whether consciousness is produced by matter or whether the brain is a filter for consciousness that may be more like a quantum cloud for example. Apart from what I term the 'animaectomy' by science, there are many scientists who will deconstruct mind and consciousness as well. When we discuss ideas of who we are and how and what the mind, identity, self, consciousness, spirit or soul are, we must consider that there are many contested assumptions. Makari underlines the role of John Locke in the identification of the rational mind as distinct from the supernatural soul and significance of events like the madness of King George (1738-1820) and the work of Philippe Pinel (1745-1826). From this rational turn came modern ideas of mental illness and the natural mind and mechanism of treatment through therapy and rational means and part of the origin of psychiatry. However, even before we answer those questions we must identify what a 'person' or human is. The Latin term 'persona' referred to a mask worn by actors and was linked thus to the role. It grew up to be a meaningful description with Christian and even legal and linguistic concepts. Some scientists deny it and confine it to genetic identity or behaviour associated with the brain. Some distinguish metaphysical and moral personhood. The body is obviously related to the idea of a person for Wittgenstein (1889-1951).

The idea that we are like machines, as Thomas Hobbes indicated, or are actually a machine, as de la Mettrie (1709-1751) argued in *Man As Machine* (1747), is also important. If you argue for this monist, materialist viewpoint and make no distinction between mind and body then it becomes easy to anticipate an improvement of the machine by other machines or substances. This is especially so if you have concentrated on a machine-like description of the operation of humans. Thus if you say that computation is a higher intellectual skill you also indicate that machine function will appear better than humans. Hobbes believed reasoning was 'ratiocination' which was computation. If you believe that the best humans can do is some sort of computation and they are like a machine or are a machine then not only do you suggest ability to enhance this function by mechanising the human even more or enhancing them, but you admit that their natural gifts are relatively weak in comparison to machines. Boole (1815-1864) developed computation and reasoning in *Laws of Thought* (1854). Babbage (1791-1871) had sought to build a machine. Mathematical logic, ratiocination, machines and theories such as 'feedback' gave rise to the modern context. The mechanics of governors or governance informed these developments. It can be argued therefore that a notion of the human as a machine or like a machine, gives rise to a reductive idea of human abilities. Intellect is reduced to thinking and capacity to think is reduced to computation because a machine can do such things. Humans are then like certain dynamic machines that need to be governed and there needs to be intervention through feedback. A consequence is that people should be governed mathematically and by machines and are capable of being assimilated into networks. But Babbage accepted higher forces also.

There is a strange paradox here in some of the most interesting streams of spiritual and philosophical thought. From the Gnostics and Cathars to Teilhard de Chardin and Terence McKenna, there is a strong sense of the need to transcend the body and move into a spiritual and mental world. Many spiritual and philosophical traditions de-emphasise the body. The result is that certain traditions can begin to sound remarkably like the transhumanists.

Thus the Buddhist ideas of non-attachment and moving into a state above suffering, is a very clear parallel to the justifications used by transhumanists. They can claim that the use of technology to avoid death and suffering is totally consistent with a long spiritual or philosophical tradition which seeks to escape or overcome the body. The problem is that certain emphases in such traditions may then be utilised to support technical movements towards the 'technosphere' or 'technium' which ousts the human. The transhumanist movement has engaged in discussions about resurrection of the individual in another form. The parallels with Christian eschatology have been noted. Teilhard and Kurzweil point to a new tech-path to creation of a god through merger of superintelligence with humans. This has been a magic, superman wish for a while.

Gnosis refers to the long history of exploration of spirituality that can be seen in C.G. Jung (1875-1961) and traced back through many traditions to early Christianity and before. In the 20th century, gnostic texts were re-discovered which demonstrated once again a significant theological and philosophical picture of the world. The term was used in the 17th century to refer to the idea of knowledge. Such knowledge is of a revelatory nature. There are certain recurrent trends in various movements of gnosis. Some even suggest a uniform basis as part of perennial wisdom. Firstly, it is experiential and consistent with the actual sense of spirit. Secondly, gnostic traditions often have an idea of a false creation and a need to return to the transcendent realm. Thirdly, it may lead to an idea of disdain for the material in some streams. The anti-materialism which can be seen among the Cathars and others is consistent with an idea that this is a false reality.

In that sense, the perennial gnostic myth, philosophy or wisdom is remarkably consistent in conception with ideas of the 'matrix' or even that we dwell in a computer simulation. This is a false copy of the spiritual world and we have come here because of a fall in the godly world. This might also explain why people like Swedenborg (1688-1772) and many spiritual explorers could encounter higher worlds which were our true home but perceived as heavenly. Similarly, the Buddhist idea of attachment and illusion could be seen in parallel. Likewise, the Platonic notion of a place of ideal forms may be related to gnostic ideas. Similarly, the Near Death Experience (NDE) of Er related by Plato and the story of coming back into this world with a draught of forgetfulness is consistent with an idea of a true world beyond our illusory one. It is consistent with modern NDE's where people feel they are really who they actually are when projected into this remarkably consistently described place. There is a sense that such traditions and contemporary possibilities of experience resonate with the transhumanist idea of accessing an incredible kingdom through electronic and synthetic means.

Similarly, the psychedelic movement in the United States that began at the same time as 'transhumanism' and was celebrated by McKenna and Leary (1920-1996), preached a similar liberation of the mind through the technology of chemicals. Leary and McKenna celebrated the dawning, digital de-coupling potential as well. They seemed to have a similarly desperate desire to escape. They both seemed to want to be liberated by technical means and thus drop out of the world. In this sense, focus on the individual is through a mentality of wonder induced through technical or material means to deal with the mundane which was stubbornly robust to alteration beyond the altered state. Indeed, consequences of deluded and drugged dropping-out aided by the substances produced through the pharmaceutical system and often for the benefit of the military-industrial complex seemed to have the effect of pushing the pendulum in the opposite direction in the generation that came afterwards. It can be argued that tie-dye rebels of rock and roll talked a good game but were sympathetic to transhumanism.

McKenna embraced the possibilities of the computer world and internet. He foresaw life-forms emerging and seemed to celebrate that with a sense of a world forming in some quasi-biological way that seems materialist. McKenna was an admirer of *Finnegans Wake* (1939). Joyce seemed to resort to a more mental world than a spiritual one in a wonder of construction. The modern mind needed to reject the spiritual maybe because the apparatus of production allowed more experimentation of a technical nature than spiritual experience. 'Psychedelics in the Age of Intelligent Machines' was a late essay by McKenna in 1999. He argues that we are machine-like but that our individuality and creativity define us. Being interested in psychedelics and computers, he saw the clear connection as involving consciousness expansion. He described cybernetics as a 'consciousness-expanding technology.' McKenna saw the bridge between mind and machine as one of symbolic logic or mathematics. Thus thinking clearly made us intelligible towards machines through precision, specification and definition. He thought nanotechnology would bridge the distinction between psychedelics and machines.

Note that McKenna is focused on the mind and on matter. Then he is focused on function. Thinking clearly and intelligibly for him is thinking in a machine-like, computational or functional way. This approach betrays a number of things. McKenna seems uninterested in ideas of spirit which are not functional or constructed. It is mind-manipulation which is the essence of consciousness in this relationship he is analysing. He does talk of access to visionary experience through psychedelic intoxication in a way that presumes additional intervention of chemical material or extraneous substance. In this way there is the notion of a necessary enhancement to reach higher mental states of consciousness defined by their 'boundary-dissolving states.' Implicit in this is an inherent limitation that needs to be surmounted by material means. Thus we have a notion of prosthesis. This was linked for him with virtual reality.

> *"So, prosthesis for the human mind, and with the advent of virtual realities of various sorts and that kind of thing, prosthesis for the human body."*

McKenna saw print as a 'boundary-defining' medium. Associated with that was a sense that *'all governments were incredibly antiprogressive forces.'* He thought we needed a shamanic input and mentioned a visit by an angel that inspired Descartes through insight about the conquest of nature.

> *"My point there is that human progress has always depended on the whispering of alien*

minds, confrontations with the Other, probes into dimensions where imagination and chance held the winning hands; so the shaman, as paradigmatic figure, is applicable both in the aboriginal social context and in the present social context. The skywalker, the one who goes between, the one who passes outside of the tribe and then returns with memes, insights, cures, designs, glossolalia, technologies, and refertilizes the human family by this means. It's irrational, but it's how it actually happens, and it's how it's always happened and it may very well be the only way that it can happen: this cultivation of the irrational, this flirtation with the breakdown of boundaries."

McKenna then proceeds to point out the significance of automatic machine-control and AI and how they already operate our world autonomously to some extent. He saw this machine autonomy in capital, energy and ideas. Looking at machines he suggested that they were the earth's way to sublimate itself into a 'higher level of morphogenetic order.' He then suggested that computers were mutating because all complex systems do. At the same time he saw humans as defined by technologies including languages which would become obsolete. Humans were suspended between archangels and animals and would evolve. He was excited that the world was changing into an interconnected plenum where all was one. In some ways he saw us as organs of the machine and that the distinction between flesh and machine would disappear at some stage. His shaman was not a Jungian one but a mathematician who could see new relationships and geometry of the universe through 'perturbations' of consciousness. Interestingly he uses the example of flagellation which reminds me of my techbondage theory. He envisaged a 'conquest' of the galaxy by transformation of the,

> *"...human soul into a galaxy-roving vehicle via our machines and our drugs and the externalization of our souls."*

This puts McKenna into the scientific or scientism camp. He is arguably in the stream of Bernal and Wells and some of the Huxleys also. It does not seem to be consistent with the shamanic vision at all but rather the scientific vision. It is also curious that Gerald Heard (1889-1971) promoted H.G. Wells. In addition, some other members of the consciousness-expanding movement such as John C. Lilly had links within the military-industrial complex.

The issue of transhumanism creates a rift in the perennial philosophy. If perennial philosophy or wisdom is not united on the issue of whether humanity should disappear in its current state, then its existence as a coherent approach or intellectual enterprise is made unsustainable. There is a fundamental issue about what it is that we are. This raises issues of our inherent nature, free will and evolution. It seems impossible to believe that unimpeded, free, technical transformation of beings can occur without diminishing them. It is highly unlikely that individual 'morphological freedom' could be allowed for significant amounts of people. Transhumanism claims to aim for retaining the essence of people in new forms. A non-spiritual, scientific, behaviourist or reductionist view of the human would yield a very pale simulacrum as an essence. Higher forces of human nature unrecognised by science would not need to be replicated in order for the human to be allegedly preserved in its essence in a new form. Empiricists focus on senses and the tangible realm.

Spiritual evolution is about a growth or movement to higher potential. The emphasis in perennial philosophy involves recognition of the body but not in a purely physical way. Spiritual tradition reflects the idea that we are more than the apparent and tangible. If we look at early Christian Theology as informed by Neo-Platonists in the work of Eriugena, the 9th century Irish writer, we see a discussion of the essential and super-essential, the sensible and super-sensible. The idea behind such theological, philosophical exploration combined with legendary, mythic, artistic pursuits, continued in recent times by parapsychology and post-materialist science reflects a much wider being of humanity. Without a conception or practice of the individual that accommodates spiritual, numinous, noetic and imaginal realms, our mode of being is radically impoverished. Disenchantment by dispiriting or consciousness-denial disillusions the person whether they are conscious of it or not. The idea is reflected in the desert scene in the *Gospel of Matthew* between Jesus and Satan. Satan shows the kingdoms of the earth and offers them on condition that Jesus worships him. Rejecting them is denying priority of the material over power of the spirit. This is not denying of physical reality in existence but recognition of the superiority of spirit. There is also the idea in the Bible in *Ephesians* 6:12 that there is an extra struggle that is not merely against flesh and blood but against powers of darkness manifest in this world and beyond, against the rulers. If one has difficulty in identifying with such spiritual discourse, they might think in terms of dark, destructive forces. The unadulterated, primary desire of power over others without other redeeming motive is a dark force represented as the Ring by Tolkien (1892-1973). When an opportunity is provided by a system of technology to gain enormous power then we can expect potentates to pursue such possibilities. Where pursuit of power can promise some advantage or benefit to people, it is easy to obscure costs. It is particularly easy to hide costs when you control the system of propaganda that promotes actual, supposed or superficial benefits. Supposed benefits that may be outweighed by costs and advantages that accrue will be but a sliver of those which accumulate for the system controllers. Accumulation reflects accelerating momentum from network effects.

A relevant school of 'personalism' is associated with Scheler, Mounier (1905-1950), Charbonneau, Ellul and Wojtyla. This philosophy diverges but seeks to celebrate the basic person that is not an object. From such a base can be constructed an idea that conserves humanity and personhood based on love as Pitirim Sorokin (1889-1968) argued. Love is critical and presupposes reason. If hate motivates reason then it will result in evil. So there is an *a priori* nature to love in a conception of personhood and the individual. Machines and instruments are unable to experience such qualities, incapable of comprehension and unintelligible in machine terms. These qualities are lost through reduced, deconstructed and disintegrated ideas of selfhood. We must be de-personalised and de-humanised to succumb to the illusion of mere material existence and existential angst must be then enlivened by a fantasy of phantasms through phantoms.

We are individual persons with constructions of self and a real essence that is a spiritual being or entity founded in an ideal form of consciousness whose existence is love and creativity. Methods of knowing, qualities and phenomena come from this base. Our ego, experience and contributions or service come from this ground of being. From our being comes recognition of other beings through their own consciousness and identity as spiritual beings. This is the essence of spirituality and mystery of life that gives rise to mysticism. When we take one projection from the ground of being such as rational thinking and the tool of reason and confuse them with the Real in ourselves then we mistake reflection of the thing for the seer. Out of this delusion comes the fantasy of construction that has confused the avatar form for the creative inspiration. When body and brain is perceived to be the sum totality of the person due to its undeniable materiality despite scientific inability to explain mind successfully, the spirit is ignored. That which we believe is the essence will be mechanically reproduced as a simulacrum and the boy becomes a wooden puppet.

Are you just a human resource? If you are, who are you a resource for? The transhumanist movement plays with ideas of spirituality because it could be a new religion or maybe a pseudo or anti-religion, perhaps the simulacrum of the incarnation. The dis-incarnation might allow an unholy spirit of AI become a new created God.

This is the ultimate hubris of Daedalus mixed with the delusion of Narcissus and fused with the failures of Prometheus leading to a digital or quantum King Midas. Philosophical streams like personalism seem to answer some questions raised. Personalism suggests holism and indicates a method of phenomenology which respects the subjective aspect that is critical in mysticism, spirituality and perennial philosophy. Personalism makes people masters over technology and not slaves. It provides a coherent protection for legal rights and proposes a respectful relationship with the environment. Furthermore, it is the natural state of the human mind unclouded by confusions caused by philosophies of power very distant from love of wisdom. Brentano was a very influential thinker with many illustrious students. He saw human consciousness as having a divine element and an interpersonal one. Personalism for Wojtyla was not based in the purely physical world or ideal world but in the person's moral centre where they chose between the right way and the wrong way.

We are people who constitute a whole entity or organism, possessing personhood, consciousness and vital energy that is not reducible to mere mechanism with an essential dignity deriving therefrom and from its recognition. We are embodied and despite limitations that is still a great and wonderful mystery defying reduction, despite our enhanced knowledge. Being individual persons we are endowed with a great fund of potential inscribed in a cumulative way through complex means to manifest in our free will and intentionality. Our essence is essentially sovereign and for many would be part of a greater or divine consciousness shared with others and having capacity to grow and expand. While many now deny their spiritual consciousness or spirit, this is the essence of humanity and seer within that can transcend the mere persona or mask that is largely socially constructed. Nevertheless, insofar as certain aspects of our outward persona are mechanisms or devices to interact with the world, we must accept ourselves and not allow our essence to be reduced through deconstruction of elements of our social person. There is an essential trinity of us that is one thing. If we are to get out of our mind we should return to spirit. Even for Hobbes, the divine is incorporeal. Consider also that the idea of spirits, unseen forces and action-at-a-distance is also the basis of magic.

While Bernal, Teilhard de Chardin and McKenna advocate boundary-dissolution, a contrary tradition in spirituality and indeed biology requires the boundary. William Blake always identified the significance of the boundary line in art. He was a formidable mystic who understood that despite our connection to everything, the nature of our existence is in the boundaries of what we are and that the imagination was the realm in which we could experience this unity from our own bounded space. That bounded space is being transformed into bondage through prosthesis that will breach physical boundary in its reaching. The problem is that potential access to the transcendental is prevented by premature prostration to a superior, physical force. Likewise, the work of Brentano, Husserl (1859-1938), Goldstein, Anders and Arendt is relevant when considering this yielding of personhood. Before we sacrifice or surrender personhood to the ideological activists in transhumanism and posthumanism, let us at least know what we give up.

There is a huge paradox about boundaries. Many are intent on eradicating existing ones. However, it is not that such boundary-challenge will be against all boundaries. Volume thereof may increase through eradicating some. At a time of erasure of national boundaries necessary to achieve a level of globalisation to make global governance inevitable, the human is living in a confined way subject to more pervasive and intrusive boundaries than they have ever experienced hitherto in history. Similarly, desire to disregard the boundary of individual sovereignty, selfhood and free will, anticipates new frontiers to the potential of personhood. Controllers of the Machine will gladly sacrifice our personhood for the promise of forced, mass predictability. Such a policy will be perfected through oscillation between the stick of fear and the carrot of fun. You will be promised a paradise in the material world and find a hell on earth. You will be programmed to sacrifice your bodily and spiritual sovereignty. You will be sold a Trojan horse with prosthesis. You will find a white elephant. You will get baited-and-switched. Remember that the illusionist is often successful. The salesperson is presentable and persuasive. Hypnosis works. Free people cannot be tolerated by people with tyrannous intent. Watch how some who sell prosthesis and enhancement will do so as prophets of liberation, as a pseudo-spiritual policy whose presumptions will be more of the materialist, confining agenda they ostensibly oppose. The true sense of boundary-breaking is exploration within, on the perennial, universal, mystical and spiritual journey that enables our consciousness combine with universal consciousness. It is clear that as some erase boundaries as irrelevant they erect them instead in our daily lives.

It is a mistake to assume that people interested in esoteric, mystical or spiritual matters share similar ideas about the basic conceptions they deal with. Buddhism is often defined by a type of reductionism of the person to a collection of aggregates without finding a unified entity or person behind it. Although there are streams within Buddhism that suggest, or seem to require, some unifying person, many other analyses are not so dependent. Such approaches paradoxically may readily support the analysis underlying posthumanism and transhumanism. Similarly, the gnostic approach of some groups throughout history and their hostility to the body and sense of imprisonment therein, could be employed to support transhumanism. Rising political movements such as Prometheanism seem to reserve personhood to those individuals that deserve it usually at the expense of others. It becomes difficult to see how practical or theoretical opposition to tyranny can be mounted without some underlying conception of the reality of people who claim protection or respect for their dignity. It is not only in the tradition of opposition to abuse of power such as manifested by Václav Havel (1936-2011) in Europe but also perhaps in Gandhi and Aurobindo and other activism that could be described as consistent with personalist thought. Both Indian leaders have a transpersonalist element. The transpersonal is the connection with our higher consciousness. However, transpersonal psychology begins in the person as does attachment to universal principles such as truth. Aurobindo criticised the idea of 'supermanhood' of the type which might be sought in a search for magical mastery of the material world and he rather underlined instead the critical need for spiritual evolution.

Technique and Magic

"In ecstasy, personality must issue from itself, but in issuing from itself remain itself."
Nikolai Berdyaev
Slavery and Freedom (1940)

The magician John Dee (1527-1609) the original '007,' conceived the British Empire and probably invoked it with a magical ceremony. W.B. Yeats (1865-1939) may have had a similar approach to Irish independence. Magic is more significant in society than people realise. Its practice is widespread back to the origins of humankind. It is secret sometimes and also studied but its nature is elusive and contested. Magic in the public mind is often about illusion and stage conjuring. Such tricks may function through mechanical, chemical, sleight-of-hand means utilising confusion, distraction and misdirection as key instruments. The illusionist may make us perceive that things happened that did not or that things that did not happen happened. Magicians play with our minds, senses, psychology and perceptions of reality using skills of deception. Skills of magicians are applicable also in warfare and marketing. Practice and philosophy of sales is primarily a peddling of illusion and creation of deep desire. So the illusionist and stage magician share some characteristics. Remember all the stories where magicians turned someone into something. Turning someone into something or locking them into some state has been a ubiquitous theme in legend and fable. The sense that we can be 'turned into' something persists in transhumanism.

Magic may be used to describe a sense of mystery, that which is inexplicable and inspirational. In that sense we see the word used in work of imaginative therapists. This general sense of magic may be employed to describe that which is mysterious and mystical. There is a close connection with the distinguishable field of mysticism. Mysticism is a distinct term that can cover a whole range of experiences which are often associated with spiritual evolution. They may involve mystical events, practices or altered states that can relate to specific stages of psychological development. Maslow (1908-1970) related such experiences to self-actualisation. This built on the prior work of Goldstein, a source in personalism.

From an anthropological perspective Malinowski (1884-1942) studied magic in the context of his field studies. He believed that magic was where people used ritual or ideas beyond limits of knowledge to attempt to exert control over the unpredictable. In essays such as 'Magic, Science and Religion' (1925) he explains how you can prepare for the sea but magic might help with the uncontrollable. Magic is a means to an end while religion has values and ends. Malinowski rejected James Frazer's identification of magic as a forerunner to science by looking at the role of 'sympathy' in magic. Perhaps quantum theory makes those connections seem more reasonable. Magic is a ritual to deal psychologically with the unknown. Agreeing with Frazer (1854-1951), he identified how magic is important in the community and is often associated with figures who have authority because they have power over plants or nature. Sorcery was often used by people who were associated with power and it was thus conservative or could be purchased by those who had power. Religion emerges from the tragedy of existence and conflict between plans and reality. It could be argued that magic prefigures science and also indicates a model for how scientists may derive social status in society including something like sorcery. Others have made the link between science and sorcery.

Mystical states were studied by William James in his classic work *Varieties of Religious Experience: A Study in Human Nature* (1902). Whether magic came before mysticism or vice versa is a chicken and egg question. It is better to see two intersecting circles. Nevertheless, there are some concepts that may distinguish the two kingdoms. Mysticism generally refers to an inner state of experience that has elements unbidden and organic and is not calculated to operate on others directly. The mystic will begin to affect the world as a result of their experience. Magic is associated with marshalling forces often through ritual activity that can operate on oneself or on others at a distance. In the Renaissance period, distinctions were less clear in Europe and it was not until after the Reformation that a distinctively Western form of magic supposedly grew up. It was informed by the Kabbalah that had flowered in the Holy Land, Spain and the South of France. It was influenced by Moorish magic through occupation of Spain. Hermetic magic re-discovered was re-interpreted in Florence and brought a stream of Egyptian magic into the West. However, indigenous traditions in Europe from Druids and the Nordic countries survived in folk practices. All countries had magical practices around the world and this was the basis of some of the early studies such as *The Golden Bough* (1890). This book posited an over-simplified progression from magic to science.

Frances Yates (1899-1981) identifies three streams of magic in Renaissance Europe, natural magic, ceremonial magic and divine magic. The latter category was more akin to mysticism to some extent. The first category is more consistent with healing practices and practical folk knowledge. The second category refers to the context of ceremonial solicitation of spirits. Conjuring is the type of magic that was utilised to summon spirits using magical incantations and spells to serve then summon to achieve their wishes. This category of ceremonial magic is the sort that may drift into the activities that are regarded sometimes as nefarious. Where ritual is used especially in conjunction with spirits to inflict pain or suffering then it is regarded as black magic by some, although such labels are disputed and rejected by others. There is a general category of sorcery and necromancy which is frowned upon by religious and secular authorities for centuries which is characterised in accordance with the summoning or evocation of spirits and especially demons and the sometimes separate but related feature of deliberate infliction of harm on others. The harm may be more perceived or constructed than intended but one cannot deny the phenomena of deliberate attempts to utilise ritual and practices to achieve bad outcomes for other people. We can quibble and debate for a long time and there are many fine books on magic and many discussions and constant re-interpretations. Nevertheless, there is a type of magic that involves a deliberate use of will to achieve power over other people. Aleister Crowley (1875-1947) defined magic in terms of the will. Will was a crucial force for Nietzsche. Will directed to power using instruments to affect and confuse people is really magic.

The idea of a mental approach to the unseen world to achieve practical results is often related to the left-hand path approach to spirituality. Spirituality may not be involved at all if magic is merely perceived as a mental force operating in a quasi-scientific fashion. Magic and science are sometimes closer than mysticism and science. Dean Radin identified magical powers more with the mystical tradition in his book *Real Magic: Ancient Wisdom, Modern Science and a Guide to the Secret Power of the Universe* (2018). When we examine use of magic as a curse or to achieve selfish goals for self-aggrandisement, then we are in a zone distinct from classic mysticism which ultimately aims to approach highest consciousness as an experience and even then communicate with the Divine. The connection between magic and science has been examined by Frazer, Warburg (1866-1929), Thorndike (1882-1965), Yates and Walker (1914-1985) and others. Magic and science were linked by operating on nature and experiments. Alchemy is closely related to chemistry and principles associated with sympathy were proto-scientific. Observation was important in both. Forces were studied in both. It was only in the last couple of centuries that science sought to reverse purify itself and alter its birth certificate to suggest some universal, imaginary identity. The Warburg Institute studies the image in human history which is closely related to magic and its preoccupation with symbol and image. There is great debate here and writers such as Yates have been corrected and criticised. Magic and science have always been related. This is why Faust was so important to many writers. Some say Western culture is Faustian and destructive.

Use of scientific knowledge and know-how to construct the tech-infrastructure of our technological governance has already combined with use of technique as applied through means of manipulation of masses without modification or inherent ethical counter-weight in science. This sounds suspiciously like the impulse towards dark magic. Many are entranced by technology, rendered dependent and will suffer through succumbing to it. The strategy pursued in Europe from the Enlightenment, Reformation, Scientific Revolution and permeating the industrial, colonial age, manifesting in a piratical usurpation of places and people has brought us to this security, pharmaceutical, military-industrial complex. The tradition of magic closely related to science has a dark dimension of egoism. The Promethean, Satanic, Faustian approach to 'supermanhood' by Nietzsche is an attitude that promotes a certain instrumental view of magic and science. The sense of techno-power might be indicated in the trajectory, for example, from Crowley to Jack Parsons (1914-1952) from magical instruments to rockets. Some transhumanists have been very interested in the occult. Hollywood is clearly engaged in a process of propaganda manipulation. The overwhelming hypnotic power of film and infrastructure of stardom allows a process of promoting messages that are never purely artistic expressions or stories calculated to enlighten the viewers but represent propositions producers wish to literally project onto their audience. You do not have to be a conservative to notice that the level of sex and violence are consistent with a warped fantasy of Enlightenment that goes back to the Marquis de Sade (1740-1814). In an essay entitled 'The Imp of the Perverse' (1845) Edgar

Allan Poe (1809-1849) talks about an attitude or force ignored by people who study the brain. Despite claims to benign reason there is a propensity to the unreasonable, unpredictable, pointless, destructive or negative.

> *"I am not more certain that I breathe, than that the assurance of the wrong or error of any action is often the one unconquerable force which impels us, and alone impels us to its prosecution. Nor will this overwhelming tendency to do wrong for the wrong's sake, admit of analysis, or resolution into ulterior elements. It is a radical, a primitive impulse-elementary."*

The superstitious or supernatural influence in the origin of science and imp of the perverse combined with the Faustian impulse imply that the veneer of rationality and reason in a society of scientism, technique and worship of technology hides a darker force. This is manipulative, tending to control and destruction. Science assists in construction of complexes that do not serve humanity. They will prize curiosity and power over responsibility, restraint and ethics. They will pontificate about benefits of science and deny costs. They will ignore externalities. We must forget about all those experiments around the world that were mean, nasty unethical or disgusting. We must presume scientists are never motivated by personal interest, power or a desire to contribute to negative institutions that control us. The tendency to black magic or the idea that one may exert a negative or destructive power through tools is akin to reckless science.

Underneath all this is an issue about who we are and how we see ourselves. If we recognise that we are of the most magnificent consciousness, created or evolved, of the greatest force of transformation in the universe, save itself, then we must respect that power we possess. It is manifest in our person. There is a distinction made by some personalists between personhood and individualism. The implication is that the latter may be more about an atomised part of a collective capable of alienation. Personalism looks at the inherent dignity of the person as a microcosm. In doing so, it is not reductive. It may presume that the person is part of Divine consciousness and often that such consciousness is within. That is different from saying that you are God. A puddle or pool is not a great lake though the substance is the same. Magical tendencies may become about power and individuals divorced from their wider community. Practice then may not involve self-enlightenment but self-aggrandisement. The ultimate aim of some magic may be to attain power through the exaltation of ego, ritual and concentration. Such a magical attitude parallels a scientific tendency that is often denied. Power over other people, nature and natures, is seductive to many. The Faustian pursuit of knowledge and power is the death knell of our society. People will go to hell for power over you. That includes some scientists. Jacques Ellul also identified this connection between magic and science. It is interesting that a type of magical thinking that is pseudo-scientific ironically drawing from the 'humanities' ultimately aims to promote technique and science while ostensibly attacking the human and reason. Magic is relevant to the transhumanism question thus in a number of ways.

(1) Magic is about study of invisible forces.

(2) Magic can be cut to exclude mysticism to focus on material instruments for control.

(3) Magic is a precursor of scientific approaches and scientism may ultimately regress thereto.

(4) Sorcery is related to undue love of control, power, transformative potions, altered states, instruments and a search for immortality.

(5) Magic is associated with story, shape-shifting and physical transformation or alteration.

(6) Magic is often about control of people and subjectivity, ritual, rules and regulation.

(7) Magic uses power of images, spells, symbols, confusion, illusion, inversion or spectacle.

(8) Magic is associated with objects, charms, statues, automata, golems or artificial beings.

(9) Magic is associated with summoning entities.

(10) Magical ritual is related to secrecy, technique, rigid codes, intelligence and imaginal fascination.

(11) Magic is often identified with hi-technology.

(12) Magical techniques may be employed to manage mass media and manipulate reality.

We may be mesmerised to supposed inescapable merger with machines, more magical than rational. It goes against the significance of persons in natural and 'Divine Light' magic and may have more in common with ceremonial magic. Many forms of magic re-affirm the natural and human and recognise the spiritual. We must look at the whole picture and not focus on a few pixels packaged by propaganda machines camouflaging purpose. We are prey of machines marshalled by the power of a mono-tech culture dedicated to control of us through data. Whether inherent in technology, opportunistic or conspiratorial, it does not matter if we are now going to be managed by machines until we merge to serve them. That we can be tricked is part of the stage-show. The magician knows how to impress us with gimmicks manipulating attention to master our mind. Entertainment promotes the system that controls us, distracting us from plans for our management. Futile 'friends' are part of the incessant takeover by the technosphere. Some messages are embedded in superficial text to speak subliminally to our subconscious. Some are in plain sight but we fail to contextualise them. Now the magic of our imagination, mythic and imaginal world of collective unconsciousness is shaped by media who project into that enormous cavern gaudy signs of things it can sell and the story of success of the quantifiable. Shape-shifting from old humans to something else will be sorcery to subjects. That Jeffrey Epstein was a transhumanist with pseudo-scientific, eugenic views might be relevant. We are mesmerised by machines, remotely-controlled and trapped. We are caught in a spell, entranced and ready to be imprisoned or transformed.

It is worth noting that Kabbalah informed the Western magical tradition very significantly. It is also worth noting that large elements are probably deleted in this process. While mysticism and magic are closely related, there is much magic that can reject identification with higher consciousness in favour of mortal power. Such magical tendencies will seek to replace the mystical approaches and most of all reject the right-hand path. The sources and nature of Jewish mysticism are consistent with perennial wisdom. It is not all magic that is rejected by mysticism. However the general approach from a spiritual perspective is not to actively seek power through magical instruments. Transhumanism is consistent with certain streams of sorcery and some of the people who created the conditions for it have recognised this fact. The prosthetic god that Freud indicated is a powerful one. Technique and technology can become like magical instruments for magicians if the world and life itself is presumed to be mere material to be managed by machines and humans are turned into shadows. As magic uses altars, alter egos, and altered states, technical alteration is based on stages, characters, spectacle, illusions and magic circles.

Perhaps if we reclaim magic in the natural and Divine Light sense then posthumanism can be opposed. Such a sense of magic is based on a relationship to the earth and to the rich, internal world we all possess. The implication of inevitable evolution of technique, technology, science and downstream, associated philosophy such as transhumanism and posthumanism is to become magic either in the sense of power of transformation or through creation of illusion. Technique and technology may be regarded as similar or analogous to magic. Unfortunately it may fit the worst of that tradition. Like Prospero could manage the weather, present illusions and compel conduct to achieve purposes apparently justifiable, our managers who control us with metrics and manipulation of emotions can exert power over us. Exertion of will in the world to achieve results is magic as Crowley identified. Like those subject to Prospero's powers, we are susceptible to being enslaved, entrapped and imprisoned in accordance with the will of others with whatever wonderfully potent and predictable wands they possess. Magic helps partly explain the origin and nature of science and indicates a mentality interested in changing reality and having power over the material world and people. This is not a mere matter of myth, history nor speculation but a manner of explaining an almost incomprehensible interaction of forces inspired by military-industrial, pharmaceutical and security complexes. Magic seems to return with scientism. When science is applied beyond its appropriate expertise, it transforms into magic to impose will through ritual and incantations. Failure of science to regard the whole in complex systems and organisms to concentrate on the part understood skews the senses of proportion.

In the children's modern classic book *The Phantom Tollbooth* by Norton Juster (1929-2021) there is a city called Digitopolis. Milo, the main character, encounters therein the 'Mathemagician.' This city is run by numbers and maths but it lacks rhyme and reason. There is a clever archetype here of an individual who seems to represent the power and magic of the modern world and of maths and computation which lie at the basis of much of the modern world. However, there is a clear deficit produced by exclusive reliance on the quantitative. We can use such tools and instruments but we must not fail to see where they do not work. There is a stream in maths which recognises the limitations thereof and suggests how its vision can include the person. Apart from Gödel (1906-1978), we can study Cantor (1845-1918) and Riemann (1826-1866). Areas such as 'transfinite' numbers indicate how appreciation of sets, may allow us perceive a higher dimension with infinite elements. This links back to Plato and ideal forms. Such openness to possibilities of maths and quantitative methods combined with appreciation of limitations and boundaries of applicability prevents claims being made that are inappropriate and may be dangerous. The Faustian love of power of knowledge unrestrained by conceptions of the person, soul or spirit is the dangerous black magic that Manly Hall indicates. As with some magical traditions, there will be a sacrifice and this time it will be the human race itself. In summary, the term 'magic' is used in a wide, variety of contestable and often inconsistent ways in diverse disciplines and in relation to different practices. It is involved in a triangle with religion and science. Without agreement about the definition of magic it is impossible to talk sensibly about it.

Some wish to appropriate the idea of magic and some wish to deny it. The fact of wishing itself is a factor which Freud identified with magic. The reason why magic is relevant to the interpretation of our world is because some elements that appear highly rational, reasonable and logical may involve psycho-social forces which represent attachments to inferior motivations. Undoubted hi-tech power may appear magical. Its employment may be for purposes other than explained. I suggest another definition of magic thus for this context.

Magic refers to a range of practices which often involve incantations, rituals or spectacular manifestations that may appear to involve supernatural or extraordinary power over invisible forces through the employment of will to achieve results or overwhelm the imagination and have effects which may be actual, social or psychological. The term 'black magic' may change and I suggest 'destructive magic' might be better in order to concentrate on the 'will to power' and willingness to cause damage. We could define black magic as destructive magic. Thus - *destructive magic refers to the use of magic as a force of will manifest in deliberate employment of practices primarily calculated to achieve or maintain power without regard for welfare of others and often with specific intent to risk or cause damage for the purposes of that pursuit.* That seems to suggest that the neutral nature of 'technique' in a hi-tech context, as indicated by Ellul, may be transformed by opportunism into something consistent with recurrent notions of black magic. Scientocracy will also use scientism as a magical practice to flummox the public with the tools of the Western witch-doctors. Scientists also engage in ritualistic behaviour.

The work of Berdyaev argued that a split occurred with humanism in the Renaissance in the minds of people like Leonardo da Vinci. He identified da Vinci with the shift to quantifiable, mechanical creativity separate from spirit and based in secular humanity. The cut with divine consciousness led to atheism which denied that we were images of something greater. Denial or divorce from spirit creates phantasms and illusions. We must submit to higher consciousness or we become useless or destructive. We should realise human nature through spirit and organism.

He foresaw a new Middle Age. He anticipated a cycle which involved the demise of capitalism. Technical civilisation would develop until it became a diabolical, sorcery. If we distinguish between non-spiritual and spiritual humanism and identify humanism with the former then posthumanism may not seem so bad. It depends on the concept of humanism we choose. Where we use the person as a base and see them as a holograph or image of Divine consciousness or Atman to Brahman and respect the dignity of individual persons then we engage in a different enterprise than secular humanism that necessarily seeks to transcend itself because it started from a limited base. Secular humanism ignores its own spiritual location in broader consciousness and thus will begin to seek transcendence in itself and through itself whilst paradoxically removing respect for other humans. Magicians become entranced themselves in objects and processes of their craft and possibilities of power over actual or notional forces. Despite the image of mad scientists and a Promethean or Faustian tendency to creation of awful, destructive machines of death, we ignore the sinister preoccupations that drive our world.

The key figure probably may be Mirandola. He attempted to integrate diverse spiritual traditions and notably the Kabbalah into Christian thinking. Mirandola sought a synthesis of all wisdom and did not exclude any major system. He was clearly aware of the mystical and esoteric life and was essentially identifying a mystical path. He believed that proper philosophy required a comprehensive view of knowledge. This was being rejected at the time by people who merely wanted knowledge to make profit or achieve some practical use.

An approach such as Mirandola's can be said to involve magic. Magic however was not separate from mysticism. Thus theurgy and communication with angels and the higher realms was part of the spiritual approach. While it can be regarded as magic, it is better seen as part of a whole, human experience at that time. To be whole one had to have a wide picture of the world and study diverse subjects. Many people who pursued knowledge for profit or power around Mirandola were most likely uninterested in the whole, spiritual approach. Technique or method which is de-contextualised and seeks to solely find power or profit can become over-specialised and have negative impacts. If magic is approached without full spiritual comprehension suggested by the early Renaissance and is utilised as an instrument or a wand of will to achieve some effect purely associated with profit or power then it is a particular type. Such magic is generally distinguishable from any quest to get closer to the Divine. Such utilitarian, instrumental approaches to magic without any adherence to the Tao, as C.S. Lewis would put it, are very similar to technique. Posthumanism and transhumanism might involve a great magic stage trick or ritual to make humanism and ultimately the human disappear. Whether it comes from postmodernism, technical prolongation of life, cybernetics or the cyborg humankind-merger, as examined by writers like Cary Wolfe, it is difficult to not see science or an associated scientism as the unacknowledged source of all. The essence of the relevant techniques is to break things down to achieve power through concentration and to build them again in a way that suits the controller or great fabricators. It was not humanism that caused what they attack either but science.

God and Golem, Black Magic

"But this rough magic, I here abjure…"
Prospero, Shakespeare *The Tempest* (1611)

The Satanic, Ahrimanic, Faustian, Promethean and Luciferian disposition and the 'left-hand path' approach to spirituality may support the magic of technology. Black or rather destructive magic is difficult to identify but many knowledgeable students of esoteric matters would recognise some elements. First, would be the desire for personal power. Secondly, would be the disposition to use nefarious means. Thirdly, would be the willingness to inflict harm. The link between magic and science and the Empire of Scientism is a strong but neglected one. Many people assume that magic represents stage illusion. Stage illusionists like Houdini were also significant in the exposure of fraud in relation to spiritualism. The magic of the stage type becomes a force to disenchant rather than allow mystery. I support the exposure of fraud. I accept that conjuring is often a trick by way of distraction to perform an entirely explicable action that is given the appearance of supernatural force. Nevertheless, that illusionists exist and are very successful and that fraud exists does not invalidate all supernatural phenomena. C.S. Lewis in *The Abolition of Man* (1943) wrote,

> *"There is something which unites magic and applied science (technology) while separating them from the "wisdom" of earlier ages. For the wise men of old, the cardinal*

> *problem of human life was how to conform the soul to objective reality, and the solution was wisdom, self-discipline, and virtue. For the modern, the cardinal problem is how to conform reality to the wishes of man, and the solution is a technique."*

If we examine the work of scientists like Dean Radin, in books such as *Real Magic,* we see an altogether more balanced approach that looks at evidence and remains open-minded in a way consistent with the work of William James. Magic apparently shares space with miracles and extraordinary phenomena explored by spiritualists and parapsychologists. Magic also parallels stages of spiritual evolution and that which is magic is often a side-effect of mystical development. All spiritual systems indicate a series of enhanced or supernatural powers that are associated with certain levels of intense spiritual development. These powers manifest real features of consciousness that remain largely untouched such is the inescapable, incessant power of the materialist, positivism, empirical paradigm. The significance of senses in the quest for truth is clear, however conception of the senses of key empiricists such as David Hume (1711-1776) is extremely limited. Science now recognises a fuller range of senses that question paradigms about philosophies predicated on a narrow scope. The work of Donald Hoffman emphasises the role of illusion and the difficulty in relying on our senses to establish truth. Science and mathematics is telling us to be careful about what we think we perceive. This is the long tradition in mysticism.

As mentioned before, if we look to the Renaissance we see a magic dimension thereto associated with Florence and the Academy with its de Medici sponsorship. As well as integrating the translation of Greek philosophy and re-visiting and re-discovering them, there was also the influence of texts from the East and from Egypt. The Hermetic texts became important. In this milieu there was a strong, universalist tendency manifest in certain figures who were interested in a range of subjects including magic. It is important to be aware of how ideas emerged from this crucible. As suggested earlier, magic was said to be of three main sorts. Natural magic was associated with knowledge of the complex interaction of cosmic forces with particular emphasis on plants and stones and use of herbs and medicine. The highest form is what is referred to as Divine Light magic which represents a system of individual spiritual growth. The more controversial side of the triangle of magical endeavour was of ceremonial magic. This was primarily directed towards the binding of spirits and invocation and evocation of beings that were in that dimension. This tradition is long and is often associated with Solomon. This is the true commonwealth of 'conjuring.' It overlaps conceptually with processes such as exorcism. It relies on the calling of spiritual forces to compel performance of tasks from lower beings. These tasks might be as mundane as finding buried treasure.

The mentality of the ceremonial magician does not require the spiritual context that mystics seek. The mystic regards any powers that evolve as mere manifestations that should not detain or distract from the pursuit of higher forces or the light. Whether it is the 'right-hand path' in Indian tradition which seeks alignment with the higher forces and karma of the golden rule, the presence of a fundamental ethical principle is crucial. Mystics in the perennial tradition demonstrate that the unity of being brings principles of respect and compassion. In this journey, ego is minimised and humility is a starting point for the grail to receive higher inspiration. In contrast to this, the magic tradition prized ego, power and compulsion to be applied in contemporary mundane contexts with the implication that beliefs in an afterlife are not as significant as the present power. This is the story of Marlowe's *Doctor Faustus* (ca.1590). It underlies Prospero in *The Tempest*. It was perhaps softened by Goethe (1749-1832). The binding of spirits and demons is an exercise in power and compulsion consistent with the mentality of the ceremonial magician. This is not the way of the mystical path and does not fit into the Divine Light tradition. Alchemy may oscillate between the two traditions. In figures such as Marsilio Ficino (1433-1499), Pico della Mirandola and Giordano Bruno (1548-1600) we see this interest in humanism and magic. Mirandola wrote 'Oration on the Dignity of Man' (1486) which laid a basis for humanism. It was almost immediately reduced.

The traditions of magic and science overlap. Newton was described as the 'last of the magicians.' Alchemy and chemistry overlapped. The study of light emerged from theology. Arthur C. Clarke said that sufficiently developed technology is indistinguishable from magic. There is a deeper connection which is based on power. Science and technology are similar to magic in that they both seek control or power over nature or people. Thus there may be a similar attraction to magic and technology for those who enjoy wielding power. Alternatively achievement of power from the pursuit of curiosity may give rise to an undue sense of pleasure from its exercise. Unlike the right-hand path, the left-hand path may prize individuality driven by ego to exercise the will to power. Focus on short-term power and disregarding means of achievement thereof or consequences of Prospero's 'rough magic' is very similar to the Promethean quest in science. The idea of Prometheus stealing fire from the gods, supposedly for human benefit is taken as justification for taking scientific risk and suggests the nobility of their search. This trick was countered by Mary Shelley in *Frankenstein: or, The Modern Prometheus* (1818). Prometheus is often linked with the idea of the noble adversary who is Satanic. Her husband Percy Shelley (1792-1822) was very taken with the Promethean-Satanic archetype. Some see the word 'Promethean' as being the most accurate, appropriate description of the Manhattan Project. Likewise, Crowley who was a famous magician was close to Jack Parsons who was a US occultist behind the Jet Propulsion lab. This esoteric-science connection is stronger than might be initially thought. Russell Varian (1889-1959) was named after the Irish mystic AE Russell (1867-1935). There is an awareness that a certain type of magic leads to destructiveness. This was perceived by Norbert Wiener (1894-1964) 'the father of cybernetics' and explained in *God and Golem Inc.* (1964). Wiener noted a parallel between a certain literalism in the gadget-worshipping technologists in science and stories about sorcery. He suggests there is a correspondence between the desire of occult power in sorcery and certain psychological profiles in science. Manly P. Hall makes this connection most explicit in *The Secret Teachings of All Ages*.

Günther Anders wrote *The Obsolescence of Man*, Vol. 2 (1956). He indicated that mankind was driven by this Promethean creation whereby that which could be conceived was made, no matter how deleterious, even if it would destroy humanity itself. That which we laud as curiosity combined with great focus and using the might of the military-industrial complex creates multiple forces of infliction of misery and even mass annihilation. Thinkers like Anders indicate how tv dilutes our perception and we could say alchemically turns us from gold into base metals. It is interesting that people who berate the old idea of the burning-in-hell God notion are remarkably attracted to a transmuted rendering of myth into the mundane of mammon. Thus look at Gordon Ramsay. He has God in his name (as well as damn, dogma, drama, dragon and moan). He is angry and judges people. He condemns them but allows redemption and has a series called Hell's Kitchen. On tv, God has been deleted and deflated as an archetype as atheists are fond of exclaiming nonsensically 'Oh My God.' Meanwhile Satan and Lucifer are made sexy and compelling. I understand archetypes and art. My argument is not that these are serving those figures but that those representations are serving the new divine digital authority. Technology will replace the straw-man caricature of the divine by being demonstrably all-powerful, immanent, paternalistic and demanding worship. Information technology is the new IT. It polices us too. PC means police constable, personal computer and politically correct. We have bitten its apple, dwelt in its cloud, consulted its oracles and used the keys to its kingdom. Cyberspace makes manifest what was imaginal space.

The work of Mirandola was surprisingly supernatural. We would aspire to be angels. The dignity of the human was informed by mysticism, Kabbalah and other magical ideas. It seems we are rejecting humanism, which had a strong Greco-Roman, Judeo-Christian and Persian base while celebrating a sort of reduced individual or societal construction that is selfish and focused on narrow, materialist, non-spiritual evolution. Dignity was related to a sense of connection to the Divine in the way Atman is Brahman or part of fundamental consciousness. Such approaches do not succumb to false promises of scientism. Eisenhower (1890-1969) warned in his 'Farewell Address' in 1961 about the military-industrial complex and public policy being taken over by the scientific elite. This has happened. The cultish, Faustian, destructive magicians are replacing perennial wisdom. A war on 'superstition' will create a much greater superstructure to support power beyond moral restraint. The clear desire for control and power may mean that transhumanism is an arrowhead in the new technological quiver. Transhumanism will be promoted by a set of mercantile and materialist mantras that through a glass darkly invoke, evoke or conjure intellectual demons and angels and actual mechanical manifestations thereof. We are increasingly mesmerised by media incantations and magical logos such that blandishments of enhancement make our doubts vanish. Tricked, deluded, in a trance, we appear to yield to instructions and insinuations spelt out for us. In the *Harry Potter* books, Voldemort seeks power and immortality. The location of the narrative of transhumanism is in destructive magic and Nietzschean despair.

Study of magic provides a way to interpret the world. Scholars like Brian P. Copenhaver in *Magic in Western Culture: From Antiquity to the Enlightenment* (2015) or Yuval Harari in *Jewish Magic before the Rise of Kabbalah* (2017) both indicate the complex research magic requires. Magic's archaic and anachronistic appearance initiates failure to apprehend its archetypal relevance. As Jung appreciated about magic, its primary operation is on individual consciousness before it can affect others. The more famous Yuval Noah Harari wrote an article in *The Atlantic* in October 2018 called *Why Technology Favors Tyranny*. Citing success of computers in games against humans, he points to the impact of AI on society in taking over cognitive function of much work and making people redundant. Because of advantages of 'connectivity' and 'updatability' algorithmic, centralising force displace distributed human choices more characteristic of liberal, democratic models. People lose economic and political power with a 'total surveillance model.' He suggests AI will give rise to 'digital dictatorships.' AI's centralising force suits the traditional methods of authoritarian administrations. He cites inherent failure of libertarian promise of the internet. We might add mobile phones. He concludes that we risk becoming inferior, domesticated animals as we become similar to a data-processing device. There is already a glowing, magical data-mist of potential, transforming transhumanist power. We are becoming integrated data objects. A culture of moral relativism will compare you unfavourably to tech-upgraded people and you will then be downgraded and degraded and unable to point to any coherent critique in contemporary, controlled discourse.

Claude Shannon (1916-2001) and Norbert Weiner used mathematics and theories of feedback to establish the close link between life and technical systems under the notion of cybernetics and information. This line of thinking can be seen back in 1868 in a paper by James Clerk Maxwell (1831-1879) called 'On Governors.' The model of how the world worked and how machines could be managed to control things began to be applied later to how humans and operated particularly in wartime. There is no doubt that these studies were mainly led by the military-industrial complex. Weiner became a scientific rebel when he realised that scientists were putting unlimited power into the hands of those people most untrustworthy for its use and distanced himself from these forces. He believed that scientists had to exercise self-control or else they would destroy science and the world. The idea of a technetronic society seeks to make a model which fits into a system of control to satisfy the obsession for predictability that the minds of the people who excel in creating control systems want. He blamed the 'gadget-worshippers' and scientists who went along without restraint when they were clearly aware of the potential use of their inventions. The military-industrial nexus and its obsession with machines and control-systems seems to be turning upon the body politic that created it or allowed it grow as a massive parasite that feeds on fear and incites problems through invented justifications for intervention. The military-industrial complex, identified by a US President and indicated by a brilliant scientist in Weiner, is uniting around a shared mission of supposed harmony by applying shared resources to war against the human spirit itself under the guise of just cause.

The Transhumanism Trance

"With that perspective, it then seems to me that the earth's strategy for its own salvation is through machines, and human beings are a kind of intermediary catalytic step in the rarefaction of the earth."

Terence McKenna
'Psychedelics in the Age of
Intelligent Machines' (1999)

Materialism manifested creates dependency to condition and control by division. The technosphere of technique and technology conditions the thoughts of all that operate within it in a number of ways. Firstly, the *network* of technologies that control our lives and make the society of spectacle and surveillance come to be, is a manifestation of materialism and empire. Secondly, the network of technologies *creates dependency* through pervasive media dominance. In that sense, as McLuhan says – 'the media is the message.' Hi-tech infrastructure creates a system that can send signals and signs that determine perception of subjects. Thirdly, content is calculated *to condition* people to its wishes and reinforce the message that gave rise to technique and technology. Thus projecting ideas of materialism, progress and science predominates. Part of this process involves assault on spirituality and values or qualities that are non-quantifiable. Fourthly, people must be *distracted*, diverted or held in attention from reality of tech-nihilism and despair. Entertainment is the holding of attention. This is necessary in order to persuade and programme people to the propaganda, products and services. Diversion is also necessary to prevent despair caused by tech-anomie. Fifthly, the spiral of technological imagery and signs creates a *trance* that may describe the state of humankind subject to the cloud or fog. Techniques that operate to manipulate us to do what we might not are a type of black magic, according to certain occultists.

The network of control that creates dependency and conditions us through distraction creates a sort of trance. In this state of suggestibility we are open to propaganda. The primary objective, mantras or incantations spelt out are as follows. Science and technology are the epitome of human achievement and so should dictate and control all existence without question to its authority. The tool is reason and it becomes an end and not merely a means. Associated messages suggest that scientists should be the new priesthood of this one true religion and all other religions or affiliations not in the scientific apparatus should be eliminated. Anything done in pursuit of highest values of curiosity and creativity is justified and anything seeking to contain or constrain this impulse is backward. The message about humanity is programmed regularly and implanted through media. As science has become predatory, it aims to destroy opposition which comes from the prey of humanity. The spell involves a process of making humans vanish. This is done by claiming humans are illusory invented constructions less than the sum of their parts, without selfhood, personhood, free will, spirit or mind. They are merely parts, neurons, bits, data collections to be ordered. Now it is not just the message that can be implanted but the means or technique of control which is the end. Their end will be our End.

We live increasingly in a hi-tech induced trance state. We are subjected to systematic susurrations of surrender. The idea of a trance, dream or a spell is recurrent and ubiquitous in literature and everywhere in every-day life. It is the state in which we spend a large part of our lives every night. It is the theatre of the imaginal world. It is the condition of daydreams. It may be that the subject is put into a coma or under a spell. We may think of a young maiden in a deep sleep. We may hear of a spirit confined in a tree. We may be told of a man who falls asleep and wakes up in a different world. We may read about people enchanted by supernatural beings. We may listen to the stories of people abducted by aliens. We may recall legends where peoples were put into some altered state. We may know that some were bewitched, entranced, others hypnotised or lured, perhaps fascinated. We may read of Mesmer (1734-1815) and Erickson (1901-1980) and how techniques are effective to lull someone into an altered state. We see people who achieved altered states with psychedelics. We experience states of intoxication. We see spinning and drumming being used by shamans to enter trances. We know that science has studied altered states. We understand about the power of the anaesthetic. We witness and practice the power of concentration. We commonly experience meditation and prayer and may be transported beyond the here-and-now. We may encounter mystical states. We watch tv and screens and enter into imaginary worlds that profoundly impact on us. We are moved by messages. We are induced to act by media. Yet we fail to see the possibility that we are entranced. We are entranced so we can be programmed and our software hacked and then our hardware altered too.

The stream of mesmerism, hypnosis and altered states is the fundamental methodology of many shamanic and spiritual traditions. Mystics and spiritual leaders have long used darkness, drumming, disorientation and drugs to achieve altered states. Such states can be facilitated in others, particularly during initiations and celebrations. The sweat lodge may use heat, drumming, darkness and herbs, for example. Plant medicines have also been used to alter states but not in such a pervasive way as industrial society is suggesting. The use of certain techniques manages access to parts of the brain or consciousness paradoxically through suppressing certain functions therein. The unconscious, subconscious and imaginal world is often accessed in these states such as hypnagogia when we fall asleep. In addition, Colin Wilson (1931-2013) talks of the robot inside us. This refers to the inherent governing aspect in human physiology that is unthinking in operation in many circumstances. We can walk and have a conversation at the same time. The problem is that these states can be hacked by hypnotism, propaganda or magic light. Advertisers, sales, marketing people, PR firms, behaviourists, psychologists and so on operate on us. Influencers affect us. As early observers such as Anders indicated, tv can manipulate mass humankind. This ability has been used as if by magicians, sorcerers and others to control and operate on people.

We fail to see that we are living in a fake reality. We fail to see that we are hypnotised and mesmerised. We fail to see that we are conditioned or induced. We fail to see that we have been programmed even when it is called a programme. We fail to see that we are remotely-controlled when we use our remote control even when they tell us.

We fail to appreciate that algorithms developed to learn about how to manipulate our behaviour manipulate our behaviour. We fail to recognise that the rote, robotic, ritual repeated actions are imprinted on us in print and other media. We are entranced and in trance. We are beguiled and bewildered. We are confused and conditioned. We are induced and instructed. We are formed and informed. We are broken and re-assembled. We become dependent and less resilient. Then with supreme irony, those who professionally promote the actual deconstruction and reconstruction of reality by creating a fake one distract us with their theories that suggest that the world beyond their false picture is a construction instead. The Stockholm Syndrome and the perception of the deceived begin to reject all reality, spiritual and physical for their construct.

There are many mystics and spiritual explorers that talk in terms of living in a dream. I don't agree with this beyond a metaphorical level. The danger is that we use this as an excuse for lazy engagement. Such approaches may be informed by science to some extent. Quantum physics is eagerly seized as evidence for spiritual truth. Within science, the work of Donald Hoffman using mathematics can support the idea that what we believe we witness is radically different from what is there. Such approaches may reinforce the idea that we exist in some unreal situation. We should accept our daily consciousness of consciousness as we normally do. Within the normal state of consciousness where we operate, I would suggest there is an apparatus of conditioning that has become very sophisticated in the mass media. The mass media through tv and the internet, in particular, allows us to be mesmerised. We are hypnotised and mesmerised.

In the 1920's awareness of the power of media to manipulate people became very evident. The work of Freud informed Lippmann (1899-1974) and Bernays (1891-1995). The ancient idea of magical ability to lure, seduce, enchant, curse, destroy, hypnotise, mesmerise, possess or otherwise overcome innocent free will has transmuted into regular occupations sanitised with scientism in marketing, public relations and behaviourism. Cybernetics represents the culmination of this system of effective control. Technology provides the media to directly access and manipulate us, as Anders explains.

The transhumanism movement involves a trance of sorts. It is promoted by the machine to further the machine and turn us into machines. The significance of the transhumanism movement is part of the overall mystique of technology and fantasy of the dominant materialist, mechanistic worldview that permeates society. This mindset creates a trance. It is not the central focus yet but as ideological momentum towards transhumanism increases, the trance will become more of a transhumanism one. We are living in a trance and from our hypnotised state we yield sovereignty. The trance in the Western and industrialised world is widespread. We are told a vision by tv. Mass media is an extension of our nervous system. This influence has been appreciated in cycles from writing to the printing press. In the 1920's there was a great focus on the use of radio and a comprehension of how potent a tool it was. The art of propaganda was refined. The idea of extending marketing approaches through mass media was refined. This idea of manipulation of the public by the invisible government who could condition the behaviour of a whole population was well understood and applied.

Some streams flowed and grew in the next generation. The war-machine racket and Atomic Age gave birth to cybernetics. Science was invigorated rather than humbled by success of its ability to exploit nature. Cybernetics incorporated insights through probing by techniques of behaviourism to increase power of the materialist view. That model informed DNA and gene research. Theodore Roszak indicated how the Great Clockmaker became the great Cosmic Programmer. He argues that information becomes the secret of life and not least facilitated merchandising. Failing to see the shadow side of powerful forces unleashed, scientific technique seeks to use control by analysis to minimise elements of unruliness in human nature. Predictable humans make management easier. An apparatus of opinion-making in the propaganda model uses psychological insights to perpetuate itself through scientism. The Enlightenment belief that science was the one true religion entitled to bind us with its worldview, mixed with other forces so a general disposition of a scientism-worldview emerged. Dominance of this scientific tech worldview is because it performs very well in some functions that are awesome. The tech-worldview can become scientism when it exaggerates efficacy, ignores costs, engages in excessive risks and exclaims exclusivity. It becomes an ideology when employed exclusively, unconstrained.

There are key elements of this total tech-worldview.

(a) *Materialism*:
Matter is all that matters. The idea of exclusive focus on materiality dominates.

(b) *Machine*:
The idea that we are a machine, biological computer, clockwork or mechanism reflects this. We then become regarded as inferior machines to be upgraded and controlled by superior machines.

(c) *Information-processing*:
Focus on information is a basis of control. While a very useful one used appropriately, we must be careful not to be wholly conditioned thereby and reduced to mere sets of data.

(d) *Control*:
Technology and technique are about mechanical control and governance. We become focused on control and subject people to constant surveillance.

(e) *Metrics*:
Measurement and quantification becomes the primary method of gaining truth.

(f) *Governance*:
The ability to control matter and people gained through science and cybernetics creates momentum to make scientific governors of the world.

(g) *Power Concentration*:
Convergence through digitisation extends to commerce and governance.

(h) *Post-biology*:
Humans can be improved because biology can be translated into information with other materials.

(i) *Progress by prosthesis*:
Adding things mechanical or quantifiable, is always seen to be better and an improvement.

Out of that worldview comes a few, specific objectives, benefits or claims that seek to make an attractive transhumanist package of claims overall as enticement.

(1) Transhumanism reduces suffering and sickness.

(2) Transhumanism can extend life.

(3) Transhumanism tends to immortality.

(4) Transhumanism allows re-coding of humans.

(5) Transhumanism allows escape from the body.

(6) Transhumanism gives superpowers.

This leads to the suggestion implied in the prefix, that transhumanism is a transitional stage until we move fully away from the human. Note that the claims and the achievements of transhumanism do not match. It is a pig in a poke. But such technology will allow control.

Another way of conceiving it involves a spectrum with humans at one end and machines at the other and a merger in between. The transhumanist tendency is to push the human along that spectrum and ultimately the posthuman stage means we have departed the humanity sector. Bernal envisaged the replacement of biology with mechanical means. Arthur C. Clarke anticipated merger of humans and machine. This is indicated by Harari in a move from homo sapiens to *Sapiens: A Brief History of Humankind* (2011). It might be seen in the shift from consciousness as human-focused to an idea of conscious agents which could include other sorts of cyborgs and robots consistent with Hoffman's view. It represents creation not evolution. Posthumanism is ostensibly a separate stream which represents an attack on a range of Western constructs but it merges with transhumanism.

The trance refers to the condition caused by massive investment in media messages by people with means and machines to make it. The media fulfil a role. It is similar in capitalist and communist countries. Both celebrate materialist and industrial models. Constant hypnosis or brainwashing by an ideology of scientism and technique of consumerism are part of materialism. The machinery uses machines and programmes to program us by hacking our emotions with our acquiescence. Messages operate on consciousness to manipulate base motives and play us like puppets with a degree of certainty about compliance and outcomes. We assume that the media in some way is our friend and the people therein are on our side. Soap operas were made to sell soap. The program then can continue as the filler. The broader political propaganda and ideology can be foisted now in much more overt ways with any pretension to other objectives sacrificed. If ads did not work advertisers would not pay for them. But now commercial forces are openly promoting political causes that somehow suit their agenda and probably not in the way that is superficially understood.

Confusion seems to be a successful technique of hypnosis. If one creates in the subject a series of statements that do not make sense, one may create a degree of desire for certainty of comprehension. After uncertainty is resolved with a certain instruction, the tendency to want to abide thereto is increased. In the ideas of Trotsky we had the idea of a 'permanent revolution.' In Mao (1893-1976) we have the idea of a deep 'cultural revolution.' We now experience a permanent cultural revolution whose sole predictability is based on its unpredictability. The purpose of this revolution is to unfold the materialist paradigm to its utter conclusion. Communist and capitalist approaches, to borrow the expression of George Galloway in a different context, are two cheeks of the one arse. Communism and capitalism are two blades of the one shears, two prongs of the one pincer. In both cases, they will lop off the spiritual head of the body politic. Capitalism and communism can be united as Deng Xiaoping (1904-1997) demonstrated. The Empire of Scientism unites ostensibly antithetical political forces in a comprehensive tech-totalitarian system of control which necessitates and involves a dispiriting of the world and humanity. Interchangeability of communism and capitalism can be seen in the origins and financing of the former, the ease in which China could use commercial forces for its ends, the movement between fascism and socialism in the early stages of the former and the movement between Trotskyism and Neo-Liberalism in the start of this century.

Let me summarise the scene to explain the inherent ideological links of transhumanism to a general use or misuse of technology, technique and science. Thus there is an ideological basis to society that is based on scientism as a technique and materialism as a dogma. This ideology is leading to the Empire of Scientism which will be based on our techbondage. The media serve this message and seek to de-stabilise us constantly as part of a permanent cultural revolution. This is the consequence of the Enlightenment and the disenchanting and dispiriting necessary to carry scientism to the nth degree. The nature of scientism and worship of technology is driven by a Faustian, devil-may-care, accelerationist, futurist or kamikaze attitude. A deliberate process or tactic of bewilderment and confusion overwhelms the senses and our critical minds to make it then receptive to deconstructing and divisive doctrines promulgated for propagandistic purposes. Science has become a major propellant of deconstruction in order to facilitate a hostile take-over of human evolution. While the left formerly provided sound critiques of capitalist society, they are essentially desirous of acceleration towards the same wall. The tech-propaganda is a pro-pagan instrumental one without the nature worship. Technology will grow the technosphere and media are the means to catch the human element in its net or web. At some stage it may become the 'Solid State Entity,' but at the moment an objective is to disconnect the individual from any traditions that may stop them embracing or rather submitting to the power of the machine system. Technology takes people away from nature and the natural. People are made afraid of natural things like the sun and other people. People's legs are of reduced importance with the car. Their memory is of reduced importance with a mobile. Their senses and intuition are less important with fingertip information. All the time we submit to algorithms and surveillance. We are made uncertain and then given certainty. We are made to embrace the standard and told we are being different, unique, strong and independent.

The people that program us and provide propaganda have a tendency to break boundaries to push with shock and awe. There is also a substitution effect. What is acceptable in meaning and practice is changed. There is a process of acculturation and alteration. Those propaganda processes are tactics that are mere parts of the overall strategy. The strategy involves full control through globetechgov via technocracy or scientocracy. Such governance must be totalitarian. As something tends towards totalitarianism it must make war on competing conceptions that might challenge it. Thus those other conceptions are eradicated or reduced. When other competing ideas are de-emphasised then messages become more persistent. Now we are not only influenced by the ads during the soap opera, but influenced by the story itself and our political world is a soap opera. *The Truman Show* (1998) is here.

There are two main characteristics of the tech-society. Firstly, there is mass diversion and secondly dispiriting. Diversion distracts from attention to higher values. Distraction is calculated to forget the psyche or spirit and yield control to behaviourists. This diversion is distinct from the process of dispiriting. The Enlightenment represented a movement not just to displace religion but to replace it. It was not just religion but the spirit and even psyche that needed to be replaced. As we forget, a new conception of the human person is conceived by addition or prosthesis presented on the basis of enhancement. Unfortunately, clear lessons from the networked, webbed society of surveillance and spectacle are that links in the various chains surround us and shape us so we become more manageable. The more manipulated and manageable we become the more digital shapers can predict and prevent us from exercising free will and behaving as individuals possessing liberty. Machine-minds will use machines to treat us as machines and ultimately transform us into machines. Transhumanism may transform some into masters but most will be slaves. You cannot help notice the way purported rulers play by different rules for their own cadre. A celebrity can jet around the world to lecture from a script about the environment not only with a straight face but a teary eye. While most were locked down or subjected to restraints others could fly to space or fill the skies with satellites. Being in a trance we can be instructed and tricked. There are many sleights of hands in the terrific, transhumanist magic show. Dismemberment in the magic box will be genuine enough. The mistake is to think that the start of becoming a machine will be big and individual. It will be small and on a mass scale.

The materialist viewpoint focuses on ratiocination or reasoning as computation and ignores higher order perceptions. Treating humans as machines to be governed as machines and by machines, is a practical, successful policy. Success of technique led its proponents to be equally mesmerised by expansion into global governance. The mathematical and machine-blinded people lead other media-blinded minions into a ditch. A certain type of brain, turned instruments into ends so the human becomes a servant to machine-logic. 'Newspeak' in *1984* or 'the fog machine' in *'One Flew Over the Cuckoo's Nest* help transform us. Instruments of the army of materialism will employ the battering-ram of transhumanism to knock down the castle-door of our physical body and even penetrate the castle-keep of our soul. Like the stoat bewilders the rabbit as part of predation, we succumb to mass hypnosis and induction through deliberate confusion. Transhumanism is a swinging pendant, a shiny object that overpowers the critical faculty, to allow promotion of the broader technique of exclusive materialist philosophy, to project the power of control through hi-tech networks. The lure of power permits promotion of a set of principles to persuade people who are already suggestible about inevitable benefits of transhumanism. The transhumanist tropes may be accompanied by tricks whose transparency is hidden by the power of propaganda from the machine which promotes it. The tech-trance allows transhumanism talk to the robot in us, to turn to the proliferating robots outside of us, to quickly turn into a robot linked to a matrix or womb of a monstrous machine that is not the golem but the god of the arch-materialist.

Steve Martin is a comedian involved in many films. His humour often has a sense of an anxious dream made manifest. He often takes some situation that gets more and more difficult to resolve and turns it into entertainment. The challenge of contexts that are without resolution is a technique of the subconscious, the deeper region of us which realises in some way that during sleep it must dramatise legion perceptions in order perhaps to allow us better manage them when we awake. The film *Planes,*

Trains and Automobiles (1987) is like a modern anxious dream transformed into art. Likewise *All of Me* (1984) involves a farce wherein an experiment with a woman's 'soul in a bowl' leads to Steve Martin being possessed by her spirit after she dies. The parapsychological study of reincarnation and possession suggest another way of looking at transhumanism. The belief in possession as a serious, real phenomenon is persistent across cultures. Transhumanism represents a potential possession insofar as the controllers of technology maintain power through the intervention process. There are examples of doctors in sperm banks who father many children. Instead of the Nobel Prize winner's children a woman gets the doctor's child. Machines may possess us also. Transhumanism is suggestive of the possibility of possession becoming a technological reality. Conditioning us to easily accept fundamental changes is made easier by the voluntary subjection of our conscious mind to entertainment in darkened cinemas from the matrix that simulates our subconscious and stimulates confusion as it programmes. Global news is a deliberate act of disturbance based on ostensible policy confusion to condition us by maintaining a sleeping, trance state. Mind-magicians are at work.

Western culture is characterised by conditioning of its people. Never before have people been subject to so much material and messages in their environment. The problem is that we do not realise when we are being effectively conditioned. If you read the lyrics of Abba songs aloud you find a degree of disillusion in text and subtext. However as they are accompanied by jolly, 'catchy' tunes you may hum along and recite the miserable mantras therein despite the dismal thoughts. Pop music makes messages. Political arguments and slogans are created by ad people. Neuro-Linguistic Programming (NLP) seems to be a common tool. A significant book on NLP is called *The Structure of Magic* (1975) by Richard Bandler and John Grinder. Note the word magic. Some call such work 'pseudo-science.' More defend it with evidence. NLP is nevertheless used as part of the apparatus of persuasion and political argument. Barack Obama successfully used slogans that sound suspiciously like that Eddie Murphy (Sherman Klump) hears on daytime tv when he watches a programme to lose weight by exercise in his role as *The Nutty Professor* (1996). All the inexplicable silly ads we see are designed to work on us. They want to hack us, on the basis that we are computers or machines. If 'pseudo-science' it would be easier to explain how it fits readily into the Empire of Scientism. I suspect such techniques are very useful behaviourist tools. However, the unveiling of the Wizard of Oz is de-mystifying. If we are to trust those scientists that claim it is pseudo-science, it would justify even more the recognition of such techniques as common policies in the Empire of Scientism. The 'black arts' of propaganda are destructive and de-constructive magic. *Logos* (Divine order) becomes commercial logos.

Finally when considering the transhumanism trance it is worth remembering the pharmaceutical industry. It is relevant to remember the apparatus of legalised drugs such as that behind opioids. Huge damage has been inflicted by such commercial systems. Apart from the death caused by pharmaceutical companies there is a persistent campaign of creation of tranquilising and pacifying drugs. Science and commerce has produced a generation in the Western world that has been stupefied with potions. These could be considered as a mass prosthesis calculated to adapt many people to the dysfunctional society wherein they are increasingly alienated, more anonymous and anxious. All these industries have vast amounts of researchers and scientists who are willing to profit therefrom without qualms able to retreat into the mechanism of it. Scientific method seems to justify anything from vivisection to vast machines producing damaging chemicals. There is also a chilling characteristic in some studies of murderers and serial killers that there is an objectification and a love of the machine and a suggestion of some correspondence with a love of control manifest in both contexts with a degree of obsessiveness. This links back to de Sade. The trance is another trope that recurs in psychological studies of these people.

We yield too easily to people in authority with white coats, clipboards, graphs and substances with long, concocted names. We trust all too easily the measured measuring. We cede our sovereignty to people often more interested in inanimate objects than real humans or other beings. Being put into an altered, mental state the public may ultimately be physically altered as well. Machines entrance us. Children become captured by moving images, screens and computer games. I do not suggest that all who promote pervasive hi-tech or transhumanism are only motivated by conscious desires to control all others. However there is clearly a strong element who do and another group who may unquestioningly assist without critical examination. The tech-tide may raise all boats and the tech-tsunami sweep all before it when it builds momentum. Those who genuinely believe or possess this hi-tech worldview and transhumanists, who may or may not be part of a deliberate global governance agenda, often resort to a sales pitch that involves tricks. When we are in a state of fascination, mesmerised by hi-tech, we are susceptible to messages that promote the policy of the producers thereof in the form of propaganda. Like stage magic, transhumanism is often presented as a series of manoeuvres or gestures which deflect from the essential emptiness of the illusion through deception. A set of recurrent arguments represent the Transhumanist Trick.

As I finish this, Abba returned to save the world. In the song *Don't Shut Me Down* they sing of transformation and ask you to have an open mind about them combined in 'another me' as you look bewildered at the shape and form they appear now. Will you let them enter?

> *"And now you see another me I've been reloaded."*

And,

> *"I'm like a dream within a dream that's been decoded."*

The zeitgeist mirror reveals all if you join the dots. The transhumanists are consistent perhaps with the 'new spirit.' Pop groups like U2, who have the name of a military device programmed us about the Zoo, are also brilliant pseudo-revolutionaries who release pressure and deflate the tyres of genuine opposition with hocus-pocus magic light and incantations of the system they serve. The lyrics of dislocation in Bad (1984) about heroin addicts really refer to us.

The Transhumanism Trick

"...the ability to produce and circulate rhetorically persuasive myths and narratives of a future technological transcendence is just as key to the development of the Transhumanist Humanist movement and their influence as are the actual technological advances transhumanists develop and promote."

Jenny Huberman
Transhumanism: From Ancestors to Avatars (2011)

Transhumanism promises future enhancement often on the basis of restoration while denying the reality of existing qualities. It rejects inherent human qualities and spirit for non-biological, mechanical transcendence. There are substantial dangers of potential misrepresentation. Transhumanism may be something else and proponents may have some other agenda or merely be unconscious actors therefor. Propaganda in favour of transhumanism may be promoted on the basis of an effective deception as to the purposes thereof. The psychological attitude of some proponents indicates a possible trajectory of the phenomenon. Transhumanism may be a bridgehead or Trojan horse for the technosphere or Empire of Scientism. *The incentive to promote transhumanism by some, in combination with commitment to hi-tech control and governance, is potentially driven by a desire to use it as a force to exert deep dominance over the human person.*

Transhumanism may be like 'conjuring,' which comes from the Latin to swear together, suggestive of banding together. It was mainly used in relation to invocation and evocation of spirits. Although magicians like John Dee would engage in conjuring spirits with assistance, he would seek to avoid evil spirits. There was a distinction between certain magicians who sought to bind or compel any spirit through conjuring and others who sought to deal with benign beings of the otherworld. For critics, whether religious or secular, the distinction was largely irrelevant. Nevertheless, bear in mind that procedures such as exorcism involve a direct but usually managed, sanctioned engagement with spirits. As indicated, conjuring has long been employed by magicians to achieve diverse goals. The procedures were well-documented and often performed in a supposed lineage that linked back to King Solomon. With the evolution of Catholicism, the processes were often enlisted to assist the magician. The necessity to draw on religious or spiritual traditional form, particularly from the Judeo-Christian heritage is clear in texts. In addition, the Moorish occupation of Spain for centuries had created channels of communication of Arabic magic. Even in the United States, European emigrants practiced conjuring to achieve various purposes and its presence arises in relation to notable figures such as Joseph Smith, the founder of Mormonism. Prospero in *The Tempest* exerts rough magic over spirits. King James (1566-1625) tried to describe such magic in his deadly work *Daemonologie* (1597).

Later the conjurer became synonymous with an illusionist and a more archetypal figure such as The Magician or Juggler in the tarot. In paintings such as that of Hieronymus Bosch (1450-1516) *circa* 1502, the conjurer works with a pick-pocket. The conjurer aims to distract us. The idea of conjuring also suggests the idea of conspiracy. If you conspire with someone you may swear an agreement. Conjuring is a very useful word to describe operation of the propaganda model upon us. Media serve the interests of the major forces that control society. In some sense controllers create a generalised spirit of the age and shape the zeitgeist. Constant invocation and evocation of the spirit of sadism, for example, can hardly fail to have some effects despite constant denial. Huge investment in media, merchandising, marketing, public relations and propaganda can achieve its objectives, consciously or unconsciously, cumulatively or more specifically. Media can facilitate genocide or create conditions for war. On a deeper level it can re-shape the mass by conjuring visions and by substituting intellectual changelings. In all this our pocket will be picked. All is a distraction. The futile fun transmuted into something supposedly compelling for a gullible and docile public is often calculated to distract. In addition, the idea of 'entertainment' itself as expressed in French and Spanish indicate the idea of diversion. Media diverts us and allows other agendas unfold. Meanwhile the unfolding agenda can also be insinuated into propaganda to prime and prepare us for its revelation. Media manipulators are magicians and many perform sorcery. Transhumanism is like a form of magic and is sold in such a manner.

How else is transhumanism like magic? Magic involves transforming things through exercise of will often facilitated by ceremony. Insofar as transhumanism, beyond restoration, seeks to grant enhanced powers that were never possible, it resembles magic. Insofar as transhumanism seeks to grant power hitherto unavailable through instruments or the ritual of technological enhancement of medical ceremony, it yields results. Insofar as transhumanism alters what was there before it seems magical. Insofar as transhumanism represents a general exercise of the wand of will to achieve change, it is magical. Where it could be particularly regarded as a negative form of magic would be when it was performed without consent or if the changes were initiated to achieve some nefarious, evil or destructive purpose as the primary goal. If transhumanism, for example, was imposed as a supposed enhancement but was really a mechanism to exert control over the individual, then the lure of the supposed benefit should be disregarded in view of the significant disadvantage. Just as a mobile phone has become an instrument of potential oppression, so could a notional, technical enhancement be used to subdue and oppress us. I do not accuse transhumanists of being black magicians. I do however point out that there are parallels between concepts of transhumanism and black magic, subject to the great difficulty of identifying that contested term. Even then, connotations in a more contemporary secular society are not as sinister as they were in the past. Certain occultists accept black magic and distinguish types. It is very context-dependent but hi-tech interference for occult reasons would definitely be of that type.

Transhumanism does seem to involve mind-tricks. A charitable view would link it to the positive idea of a trickster. There are indications of possible tricks in the presentations of arguments in support of transhumanism. There is something definitive about some of the arguments about transhumanism that are less statements of truth than arguments which seek to create an illusion of certainty. Sometimes the idea of a *fait accompli* can make an event happen. These are some of the potential obfuscations that seem like tricks when manifest as statements or strategies involved in the general transhumanist package-case.

(1)   Transhumanism is inevitable.

There is a strong propagandistic argument to the effect that the inevitable progress of technology must lead to fusion of humans and machine. What was stated as a policy became a prediction or prophecy until it is regarded as inevitable. Bear in mind that people like Bernal expressed their desire of this outcome and others persisted with it. There is no inevitability about transhumanism save insofar as successful prosecution of the policy is projected onto a gullible public. An argument is constantly repeated about exponential growth of technology and AI with the suggestion that it is somehow almost organic already. It fails to accredit the full range of circumstances and conditions of policy that facilitate such technological growth. That which becomes exponential does so because of the massive acceleration of policy to achieve it and not because of a magical singularity. In reality, the ideology of scientism and associated desire to create an Empire therewith, seeks to promote transhumanism, not as a consensual endeavour but as the ultimate collectivisation of humanity. Thus transhumanist fatalism or determinism must not be accepted or assumed and instead can be seen as a promoted philosophy that has a number of goals and interests involved. Transhumanism comes from human intent and the inevitable argument is more like magic.

(2)   Transhumanism is beyond our agency.

Associated with the idea of transhumanism being inevitable is an idea that it is happening of its own accord almost without our agency. The ironic suggestion of an inherent vital force in technology and technique may seek to avoid accountability and responsibility for involvement in a policy that promises to end the human race as we know it. The danger is of scientific domination by elite, global governance committed to transhumanism and associated enslavement or eradication of humans. Science is often prepared to bring into being first that which may do damage and await regulation when problems manifest. It suits the agenda of proponents of global scientific governance to suggest that tech-transformation is out of our hands. But we do have absolute agency still unless we acquiesce more or approach it as if it were magic.

(3)   Transhumanism involves inevitable, individual evolution.

Inherent in many arguments is the idea that technological development is out of our hands, inevitable and also that this somehow represents the ineluctable path of progress of human development. Thus transhumanism with AI is going to happen, because it is happening and it must and will be the next stage of our evolution. This argument is a weak one that is unfortunately perceived to be strong. It suggests that fusion of humankind and machine is the inevitable, optimum, obvious and right process of evolution. This disregards the fact that science and scientism propose to stop natural selection. Evolution would be replaced with deliberate control in the nature of eugenics and prosthesis that would represent devolution through over-concentration on ratiocination. The remainder of the body would only increasingly adapt to the technosphere further reducing compatibility with nature. The study of evolution identifies that the use of separate, exterior tools gave humans advantages as generalists that were not available to species who developed tools as integrated physiology. Having tools integrated in your body limits potential rather than enhances it.

(4) Transhumanism involves inevitable evolution of the human race.

The central idea is that evolution through transhumanism will represent evolution of the human race. There is an assumption therefore that transhumanism must apply to the race or alternatively such as would have access. The only way that the whole species could move on would be if it was done together. This suggests either compulsion or a caste system. This is likely to be the agency of transhumanism utilising mass networks to achieve low level, lowest common denominators consistent with control systems. Artificial intervention is not evolution but the end of it as we become the guinea pigs.

(5) Transhumanism is divine.

The slickest of sick tricks perpetrated by proponents of transhumanism is the proposition that this spuriously inevitable supposed evolution of humanity represents an almost divine unfolding. The suggestion is that such a linking of consciousness represents a level of evolution that rises to the divine. The arguably autocratic argument presented by Teilhard de Chardin seems to suggest that transhumanism is a way of humanity becoming divine. Such an argument takes religious concepts and transforms them into scientific terms or contexts. This is a sort of scientism that fails as science and religion but seems to serve interests of transhumanist proponents. The God-like interconnected consciousness that dispels individualism in some great sacrifice to a machine-like, collective existence is an insidious untruth that conceals a techno-totalitarian tendency. Advocates of transhumanism realise that they can use Teilhard de Chardin as a soft-focus support for their goals. They can pose as being consistent with Christianity in some enlightened way while being fundamentally opposed to it and especially its notion of the person. Allied to this use of Teilhard de Chardin is a flimsy link back to an old English translation of Dante which may suggest to some that there were some vague connections with religion. Posing as some pseudo-scientific notion of evolution, the intentional interference and thwarting thereof seeks to bring creation to the hands of individuals. This has always been the Faustian, Promethean, Satanic will. Machinery and mechanism will create omniscience, omnipresence and omnipotence.

(6)   Transhumanism will enhance our essence.

The idea of physical or performative increase implies some overall enhancement. Physical enhancement in one context may impair development in others. There are costs as well as benefits. But when we talk of consciousness and mind being uploaded or more realistically enhanced there is a really serious problem of analysis. When scientists and technocrats who are intent on creating intervention in the brain to adapt it by alleged enhancement, do not adequately comprehend consciousness or qualia and will even deny they exist, then how can we trust them to deliver improvement? If you do not properly evaluate what it is you have then how can you enhance the totality? This is the Narcissistic and stage-magician illusory element of transhumanist enhancement. It is a trick that may even fool the performer. You cannot enhance that which you do not appreciate exists. This is a Procrustean attempt to put human consciousness into a prison that is based on a very narrow and deluded picture of the brain as a poor, organic calculating machine necessarily inferior to manageable machines of more manipulable matter. The trick with the mind will be to present a very limited picture of human capability by fashioning its supposed content into the lowest common denominator as determined by notions of quantifiable performance consistent with basic functions often related to computation, perception, detection, data assimilation or directability. Then when you establish a very specific quantification of performance, enhancement will be easy to identify. Such 'enhancement' will be so regarded even if the value is dubious or it is acquired at the expense of some other function. If scientism denies our essential being it cannot enhance it. They will shoot fish in the barrel of a part of our mind. They will pull rabbits out of hats. It will be pseudo-enhancement without respect to qualities.

(7) Transhumanism is motivated by love of humanity.

The suggestion by proponents of transhumanism is that it is motivated by a basic love of humanity. This suggestion is usually supported by reference to spectacular examples that are often unrelated to transhumanism. Transhumanists often express a type of horror of the human, life and decay and a fundamental and a daft desire to design-out evolved failings thereof deriving from fundamental discontent therewith. It becomes difficult to sustain this pretend argument. Transhumanists love machines and not people. Instead of engaging with self-improvement transhumanists wish to engineer out the negative, presuming that which remains is positive while concentrating that selfish, self-centredness to the highest degree.

(8) Transhumanism being inevitable, evolution requires humans justify their existence.

There is a cheap but remarkable con-trick that scientists mastered and scientism and other chancers use effectively. This is about the burden of proof. For example, the prosecution must prove that an accused committed murder beyond reasonable doubt. That is the burden of proof. They must prove it. For nearly two centuries at least, scientific figures have shifted the burden of proof when there was something that had long been proven through experience, practical and tacit knowledge. For example, in the latter part of the 19th century there was a very successful campaign of what I call 'dispiriting.' The conception of the spirit or soul that had long been accepted as a reality was thrown out. It was not that the spirit was disproved, it was abandoned and marginalised. The role of theology in a shift away from a certain type of Calvinist thinking informed the milieu within which ideas of the free market (Adam Smith 1723-1790) and empiricism (David Hume 1711-1776) operated. Thomas H. Huxley led a campaign to banish not only God but the supernatural and the spirit. There was no necessity to throw the baby out with the bathwater. The spirit, or something like it, was accepted in most cultures. Science ruled it out in practice. This theft from the cupboard of universal culture meant that spirit disappeared from much discourse. Psychology was a replacement in the scientific stable. But the substitute is not the same thing. Psychology has a far more impoverished starting point of what is being dealt with. The subject with a spirit became more of an object with a mind. This strand continues. The trick remains. Make someone prove something they do not need to. Employers do this. They give a job and later make people justify it. It could be that individual humans will have to justify their very existence to a robot programmed to achieve some noble objective decided by AI or some technocratic dream. Already some are talking about reproduction licences.

(9) Transhumanism is a libertarian enterprise enhancing individual freedom.

If transhumanism was a pursuit that people followed of their own accord, individually or in groups, say like fencing or bridge, then it would not be a threat. However, transhumanism is proposed as tending towards the universal. It is not only that transhumanists want it for themselves but they want it for others. It tends towards a utopian application for all. This desire to project it onto others is the real concern. It betrays the real reason for transhumanism as an ideology that is part of the Empire of Scientism. Technology needs to be widespread to create networks. Networks create benefits from network effects. Without the mass of people involved there is no commercial benefit in the short term and no political or corporate control dividend in the long. Transhumanism has no democratic mandate. Many hi-tech networks have power beyond effective regulation. A low-level mass implant would make governance very easy for global rulers. This would be consistent with the world that Wells foresaw. There is always a good reason that can be found or made. Mass, low-level implants can be promoted without effective consent through dependency-creation. Then the choice is proved to be illusory. The ones who will benefit from life enhancement are the billionaire business and progressive neo-liberal technocratic class perhaps driven by some pseudo-green justification to maintain their coherence and governance. Big business threatens liberty as much as big government but many pro-marketeers are not prepared to admit this.

(10) *Transhumanism is inevitable enhancement verified by quantitative methods.*

Transhumanism is primarily based on the power of quantification of intelligence especially where it is reduced to mere ratiocination or some other measurable feature. If we find technological addition or enhancement that can be added as prosthesis to the body and there is a demonstrable increase in some function, the presumption is that the demonstrable, quantitative increase represents inevitable enhancement. However, that depends on overall utility and significance of the function and opportunity cost of technical enhancement. London taxi-drivers increase part of their brain through use of memory. If that function is outsourced to an implanted GPS system, the natural, organic enhancement of the brain is impeded and probably reduced. The fact that we can demonstrate a quantitative improvement in some function is not the end of the story.

> (11) *Transhumanism does not take away anything, especially non-existent qualia.*

When something is added, it is likely to reduce the relative importance of some other biological function. If you use a car all the time, you will not get enough exercise as walking is a key purpose of the body, for example. While compensation can occur through the additional cost of attending a gym, to add to the car costs and opportunity costs, some qualities that are diminished through outsourcing will be more subtle. When the empiricists talked about the central significance of our senses, they had a reduced idea thereof. We have not fully explored our physical, mental nor spiritual capabilities and many latent and unexplored gifts may be sacrificed or squandered through relative or actual ignorance in favour of an apparent, immediate benefit. Qualia, we are not sufficiently aware of, may be lost or forgotten and attributes that are obscured or contested may be ignored. The siddhis or spiritual gifts consistent with one idea of magic are less likely to evolve in the context of a mechanical substitute or simulacrum. Thus the possibility is that certain paths in spiritual evolution are blocked and our reliance on technology and further ensnarement in the technological tendrils are increased.

> *(12) Transhumanism is prosthesis and all is prosthesis.*

As indicated earlier, another trick is to say that all is prosthesis. If you argue that everything is prosthesis in that we have complex requirements from birth onwards in order to survive, then one can argue that more specific prosthesis involving technical intervention in or on the body is merely an incremental continuation. Such an argument proceeds from an exceptionally wide definition to cover a type of hi-tech intervention and transformation that is radically different. Thus some might suggest we equate a pencil with a neurological implant. While writing might be described as prosthesis, it is qualitatively different to compare it with neural implants. If all is regarded as prosthesis then technological addition is not regarded as anything unusual.

> *(13) Transhumanism is not artificial and if it is, artificial is good.*

Some argue that transhumanism is merely a continuance of that use of tools and things that we have used since time immemorial such that transhumanism is really 'natural' in some way. Science argued for the significance of natural selection. Now it is not so important when opportunity and will allow manipulation of nature. Then, as appropriate, it is argued that if it is artificial, artificial is good. Thus artificial intelligence and artificial selection are now somehow inherently or inevitably better than natural selection.

(14) Transhumanism is merely medical intervention.

The equation of transhumanism with mere medical intervention is a false one. Restoration to a sense of normal, physical expectations using technology is not transhumanism. Such arguments obfuscate and obscure. Dental implants, orthodontics and cosmetic surgery may go beyond restoration at times. However, even then the need for surgical and technical intervention, for example, is often a symptom of problems caused by an industrial society. The need for artificial limbs often rose in conjunction with military engagement. Technology and the scientific technosphere create the need for more technology to adapt to its needs. It is like we move beyond the illusion and the magicians having actually sawn the assistant in half demands glory for re-joining it.

(15) Transhumanism refers to all scientific, technical progress.

As mentioned already a blurring, obfuscatory argument is to elide transhumanism with the totality of scientific and technological advances, especially medical ones. The implication is that any sufficiently developed technology or medical advance is in some way transhumanist. This elusive rhetorical equiparation enables caricature or false polarisation of the argument as between transhumanists and all others. It is interesting that if any sufficiently developed technology is indistinguishable from magic, then the spurious claim that all such hi-tech advances are inherently transhumanist in some way does suggest an inherent affinity between magic and transhumanism.

> *(16) Transhumanism is a corrective antidote to racism, sexism and discrimination.*

The transhumanist community has a nexus to posthumanists such that the claim that they are righting the wrongs of historical injustice can be woven into the fabrication of their argument. A false claim to some sort of justice made possible by transhumanism conflates ideas of social and technological progress on a mistaken view of humanism. That transhumanism is correcting humanism by deconstructing it is not true. Similarly, risk-taking experimentalism with hybridity allows a godlike power without concern for the dislocated beings transhumanism could generate.

> *(17) Transhumanism is just scientific and hi-tech progress without political motivation.*

Closely related to arguments about technical inevitability is an argument of neutrality and an implied denial of political ambitions. However the idea of progress informs left and right and emergent governance will exploit concentrated, converging power to form a new coalition united by a policy of scientism. The 'manifestos,' claims, policy, campaigns of persuasion indicate the potential for political power posing as sound administration. Transhumanists now are as likely to talk about tax or universal income as they are about their real desires such is the effort to cloak the extraordinary enterprise in common garb to make it appear normal and ordinary.

(18) Transhumanism has achieved much.

A remarkable aspect of claims about transhumanism relates to implied success suggesting ownership of and involvement in great achievements by transhumanists. They suggest they are responsible for much progress in some magical way and create an air of success by association. However, chilled heads and bodies are still in the fridges. Life extension has not (as of yet) beaten natural life lengths. There is no evidence of people in general prospering yet. Instead the rich get richer and the poor more confined. The suspicion that transhumanism is really what it was declared to be in the start and will become a part of an oligarchical control system is more persuasive than the promise of a paradise of constructed beings. Space travel is not inherently transhumanist for example.

(19) Transhumanism can take suffering away.

The idea of super-happiness is founded on a plan to cut out the bad bits of humanity and thus reduce suffering. This is based on a simple unwillingness to accept humanity and will merely be a tool for total control by making zombies.

(20) Transhumanism is only about superlongevity, superhappiness and superintelligence for humans.

The supposedly benign claims of transhumanism are related to sale of the phenomenon as merely about helping people live longer, helping people avoid suffering and helping people be superintelligent. However despite the rhetoric about only helping humans, transhumanists may stick their nose into every nook and cranny. If they are talking about tax and what is wrong with religion and why genes should be edited, they are talking about a force not confined to more acceptable elements they strategically opt to focus on. Superhappiness may be based on genetic intervention that removes potential from the individual without consent in pursuit of an improbable and dubious objective. Superintelligence may be more about creating machines which ousts us and are thus not about enhancing humans. Superlongevity is a strange policy for a type of people who are of the ilk that think we are over-populated. It does not make sense that we can increase the lives of all when resources are under pressure. If such superlong lives are to happen, it will only be for the elect, the chosen few, the new technotribe that subscribes to this brave new world. This has been foreseen by the early transhumanists where the policy is merely part of the strategy for scientific domination. While some individuals do want to do things they perceive as possible, the reality is that they are serving the goals of another agenda, whether they realise it or not or care about it or not. Transhumanism is not merely about helping humans save happy time before their grave. Sincere people may believe otherwise.

The transhumanist trick allows claims to be made about a future that involves a mass sacrifice of ourselves in the present to machinery while obliterating our past. Berdyaev wrote about the movement from the organic to a machine that must organise and be organised. Not only may we get new organs, society forms organs in the body politic. In crossing into the technological cave, we must sacrifice our connection with nature and forces we do not understand yet. This mechanisation must even apply within our soul. A mechanical body and a programmed-mind will be something other than a flesh and blood, adaptive organism that is part of nature. The flesh and blood inspired by spirit can solve problems we face if the force of it is unleashed in a creative way. This was the Christian conception for him but more of a mystical type. Berdyaev pointed to a new Middle Ages following Vladimir Solovyov (1853-1900). This is consistent with some contemporary streams of thought. He posed a spiritual truth against a materialist truth. I think that the need for presenting an alternative to our present predicament is not yet as important as preventing the predictably perilous. The energy of other potential networks will arise to murmur in counter-balance to the monstrous concentration of machines and also will marshal new pragmatic and cosmopolitan possibilities eventually. In the small space between isolation and collectivisation is a channel of personalism.

Personalism as part of theology perhaps can be found in figures like Saint Theophan the Recluse (1815-1894) and figures around Kiev. He referred back to the ground of being of the Eastern Church. He considers the person of the mystic with an affective domain of the heart that is not purely will. The spiritual person is in the heart and where the Divine dwells. Freedom is there. Ouspensky (1878-1947) was influenced thereby. This personhood connects with others. Heaven is inwards. Berdyaev examined slavery in *Slavery and Freedom* (1939). Slavery was a deeper issue associated with the human person and the evolution thereof free from fear. The power of the spirit comes from this freedom. He notes that spiritual ecstasy may be slavery too. Part of the transhumanism trick is based on developing dissatisfaction consistent with Enlightenment disenchantment and discontent of a capitalist society that cultivates consumption by existential greed, lust, passivity and anxiety. Less is more sometimes. As with all the suggested sweeties, the initial phase of marketing is all smiles and bonhomie. Later you will find that such grins resemble the bared teeth of a Rottweiler. It will never be your way but always the superhighway. It is no accident that personalism became strong in places that experienced totalitarian collectivism. Berdyaev was one of a number of Russian personalists which included Vladimir Solovyov and arguably Alexandr Solzhenitsyn (1918-2008). Krakow was another centre.

Transhumanism may be the epitome in worship of hi-tech gadgetry as an instrument of control for control-freaks. We must disentangle strands which are called transhumanism such as technological-assisted restoration of bodily function. Entranced, we are susceptible to subjugation as we are sold hi-tech toys, presented dissembling with illusions, mesmerised with visions of E-topia, lost in our own confusion and dispirited by deconstruction and disassembly. It might be the Dionysian attack on form as an alchemical transmutation of the clarity of Apollonian form. But the suggested alteration through supposed augmentation is a sacrifice of the vital, alive qualities for deadly power of computation and control. The person, as advocated by personalists, cannot thrive or flourish where they are being sacrificed to, or substituted by, technology.

It is worth remembering how modern society is already disembodying as well as dispiriting. When we extend our nervous system by tools or prosthesis and insulate ourselves from nature through barriers, or fall out of relationship with forces we are adapted to align to, we become disembodied or stretched beyond elasticity or resilience. The prosthesis or addition will often be an apparently, superior substitute of some function we have inadequately developed or appreciated. Curiosity and opportunism unconcerned about practical consequences while combining with committed transhumanism and posthumanism, promises morphological 'freedom.' This will ultimately mean a breaking down of all boundaries to achieve some technical monoculture masquerading as morphological cosmopolitanism. Animal-human hybrids, cyborgs and robots will be celebrated as conscious agents but dangers and potential misery ignored in favour of cultivation of certain attributes merely because they are possible.

I do not suggest that all transhumanists are doing something bad or are motivated by negative or destructive forces. Nevertheless, there is a general, package-argument or pseudo-philosophy about transhumanism that is put forward. Such an argument seems more than likely to be part of the materialist, scientific tendency to assume governance through creation of dependency on its own technique. Putting aside benefits of technology and artificial aids for people who desire adequate or normal functions or legitimate entitlement to restoration or perceived enhancement where possible, the real agenda of technological incarceration and transformation in order to achieve control of the many by the few and for the few through exercise of power akin to magic, is achieved by false promises, creation of illusion, diversion, distraction, hooking-up and addiction of individuals to actual, overwhelming, algorithmic power in a concerted process of dependency-creation. The sense of things being out of balance, as some artists perceived and certain philosophers predicted, will be revealed for people to see that the picture of the 'real' world they were sold does not hold great promise for the person. In a trance by a spell, tricks of transhumanism bewitch the mind lulling the soul to eternal sleep. We are changed as we are sped spelt to be slept. Unduly specialised focus on tools, technique or technology reduces awareness of the implications of reliance thereon. The technician becomes focused on technical detail, opportunity and benefits. Personalists in Russia might be contrasted with the diverse 'Cosmist' movement sympathetic to transhumanism. Konstantin Tsiolkovsky (1857-1935) was a pioneer in rocketry and space exploration. He shared Bernal's interest in superhumans liberated from former forms. It is a mistake to dismiss insights of those interested in transhumanism. Nevertheless, brilliant individuals may exhibit intellectual hubris which makes them servants of autocrats they may never have intended to serve. Some do not care and demonstrate cold disdain or aversion for other persons that do not share their interest in escaping their home planet because they have created a prison-zoo. Mathematics itself may be like magic and losing sight of its reflective nature reduce humanity and reality. The transhumanist agenda is promoted by tricks and illusions like those of a stage magician. Time will tell but we should not let such dream merchants enchant us by telling us who we are and should be. Transhumanism is about fabrication. It is sold by fascination. Transhumanism is fabrication by fascination. Like fantasy, we are to be made up on the basis that we are just made up in some inferior intellectual way. Transhumanism is similar to magic and is promoted sometimes by a disposition or psychology consistent with that articulated by magicians who seem focused most particularly on their own individual existence above all else, beyond even morality. If it were merely an individual pursuit it would be inoffensive enough. However, it is increasingly appearing as a thinly veiled threat and in the context of a general effort to embed technology in society such that people are dependent thereon. Then choice will soon disappear. The most plausible analysis is that transhumanism represents that which has been a consistent policy of the scientific elite preaching scientism and the new theology as part of a global scientocracy. It may also be used as a convenient cover for bio-tech gambles. That scientocracy must be totalitarian. But if it reaches that totalitarianism that technology tends to then networking the brain in a collective way may prevent existence of independent subjects as they sacrifice sovereignty and end the race for the majority while allowing the ruling class thrive in some enclave of masters of the enslaved. Transhumanism is a big ad to add on that does not add up.

THE THIRD PART

Sensing the Machine-Human Merger

That humans could be confined or dominated by magic or machines, even if administered by dullards and minions, is an ancient, recurrent perception now being actually realised. That which masquerades as libertarian will easily elide into imposed, maximum, mass constraint through minimum technological intervention presented as enhancement for some pressing purpose and administered by sorcerers' apprentices. This section seeks to examine trends noticed by certain writers. The Enlightenment materialist force aims to take over everything and sacrifice human health and happiness for a delusion of certainty and predictability using cultivated forces of fear and greed. People who cautioned us about totalitarianism also warned us about techno-totalitarianism. Opportunistic drives for global governance and a centralised policy to create a new industrial, information infrastructure should be interpreted in this context. Transhumanism is a useful precursor to prime the public. Acculturation to accept transhumanism anticipates mass mechanisation of the human body consistent with illiberal management of a global population. That we will be transformed by, and into, technology is an inevitable tendency of those incapable or unwilling to appreciate the stunning, existing, biological complexity of the natural world perhaps because of their own deconstructive, dominating attitude.

Goldilocks, ER and Lagom

"And then, there is a danger that men will feel that they are mere cogs in a vast industrial machine because it is an industrial world, and man so often becomes depersonalized; the machine becomes the end. This sense of not belonging, this sense of loneliness, characterizes modern life."

Martin Luther King
'Living under the Tensions of Modern Life'
September 1956

Fable, myth and legend from ancient times prove to be remarkably malleable as new facts of being emerge. It is clear that there is a proportionality that exists, a sense of order, balance or Tao that is appropriate for the health of individuals and communities. It may be the middle path, the narrow path, the way between the left and the right or the point of balance. Our emotions and lower instincts are dangerous as they may oust higher, more considerate and compassionate dispositions and displace informed choices. Creation of instability, imbalance, lust, greed, avarice, unreasonable expectation and disproportionate longing in individuals atomised like chickens in a factory farm, are effective tactics to unpick the real links of community that otherwise work to maintain stability. We are furthermore imbalanced by incessant assault of technical intervention into our nervous system as we are caught in the tentacles of the technosphere that creates anxiety, anomie and angst.

The essence of transhumanism and enhancement through prosthesis involves the merger of humans and machine. At one end of the scale is a small technical enhancement and at the other is a machine. Any merger is somewhere on the spectrum. At some stage the essence of humanity will be lost to something which is a cyborg or a machine, with less and less biology. It is possible that a small level of technical intervention if promoted by a centralised force might have a radical, compromising dehumanising effect. Such effects will be more likely when proponents of technical enhancement or mechanisation of the body have no strong conception of the person as a real phenomenon. When scientists doubt that the mind exists, it will be easy to ignore our free will.

Transhumanism involves our supposed enhancement through technological or mechanical means. That there will be a process of merger and evidence suggests a determined disposition exists to make this union happen and indeed to compel it. Such a fusion must ultimately result in the end of humanity as we know it. In such a case the human could become finally a machine. In addition, accumulation of technique in the technosphere means that there is a Machine, in the way conceived by E.M. Forster, Lewis Mumford or John C. Lilly. As we are increasingly forced to use devices and apps and are integrated more and more into the network, we will become part of it through our dependency. It is when we become utterly dependent thereon that we will be compelled for some supposedly good reason or other to be technologically linked through a device or an implantation. This enhancement will be compulsory. Then the supposed enhancement will be seen to be a fetter, chain or manacle to enslave us. That an instrument of containment will appear scientific or hi-tech or even beneficial does not render it any the less potentially restrictive. Nano-particles, for example, may appear unobtrusive and be sold as something that serve some other purpose. The sudden appearance of 'contact-tracing' on a collective level revealed how something sold as enhancing our lives may really represent a tool calculated to contain us. The disclosures of Snowden and Assange revealed a vast, mass apparatus of disproportionate, ubiquitous, inescapable, surveillance masquerading as security without obvious democratic base. The military-industrial complex has made the human a permanent suspect and target consistent with a coalition of hi-tech conspiracy.

The individual's human body, mind and spirit is now the battlefield location for the war on consciousness. The Empire of Scientism and technique of technology aim to colonise the human and transform it into something else. Entrancing the individual has direct effects on their attitudes and behaviour. The persuasion or propaganda proceeds with a powerful technique calculated both to programme the observer and maintain the Enlightenment, materialist environment. Applying the technique through technology has consequences which allow transmutation of emotions that may be enlisted to control people. The mantra of the trance is about imbalance. The position that needs to be attained for a lack of stability revolves around a sense of the idea of what is sufficient or what is enough or not. Associated with this is the idea that something is not good enough for us or we are not good enough. Both are techniques which create imbalance.

When I went to the US, I was surprised at the tenor of many of the ads I saw on tv. The plot was the same. At some social gathering you realised you had something wrong with you. You were treated worse and were unhappy. Realising that you had a problem you reached for the solution. Then you were happy and people loved you again. There is great power in creating a gap between what you were happy with before and what you need to be happy now. Expectations must be increased and satisfied. This technique is based on fear and de-stabilisation. The message is that you should be afraid that you will be ostracised because you are not enough. Strings of the media propaganda model stretch from the needs of the puppet-master directly to your emotions using scientific observations and technology to align you to whatever commodities or service it wants to make you pay for.

The other side of the coin is to make you believe that other people are not good enough for you. This is done by telling you how fantastic you are and what you are entitled to or deserve because you are worth it. An ex-Editor of Cosmopolitan magazine said that they used to use such leverage to encourage women to have very high expectations so they could sell products. This is based on exploiting greed and vanity. In *Aesop's Fables*, the fox coaxes the crow to lose its cheese by praising its singing voice. The consequence of appeal to greed or vanity is that the person or being loses something. That is the direct consequence. The cumulative effect is that a person may find that they have a range of aspirations effectively irreconcilable and impossible to achieve. Such aspirations come from inculcation of entitlement to access to limited resources merely based on the programming.

The culture of longing, aspiration and envy is cultivated by the presentation of the new, unexpected, shiny and necessary. Advertising works by stirring up the most base instincts. Many religious or spiritual traditions talk of renunciation, breaking of attachment, being not greedy and so on. Merchandising must reclaim the values that an immaterial, qualitative or spiritual approach largely rejects. Bigger, better, quicker, longer, faster, more convenient, tastier, softer, harder, shorter, crunchier, extra, faster still, quieter, smoother, longer-lasting, jumbo, petite, brighter, darker; all these things with music and projected happiness of someone else, paid to be so, programme us to love the comparative. If we have that 'er' then we know we have moved to a better place. Funnily enough, Er was the name of one of the possibly most celebrated Near-Death Experiencers. Er went into the next world and returned after having been dead for several days in Plato's writings. A characteristic of modern day people who have had such experiences, which I would put into the category of mystical, is that they often become less materialistic. Such experiences help people realise who they are and re-focus on spiritual values.

One theatre which facilitated transhumanism has been plastic surgery. Again, much of the incentive to change the body is to repair or restore to normal function. The growth in aesthetic surgery is different. While it may in some sense be related to a base of expectations, it is often about enhancement and something beyond that which is strictly necessary or even within normal expectations. It is now common to seek 'medical' enhancement of lips, tummy, eyelids, penis, labia, face, nose, ear, brow, or buttocks. Associated with the standard reasons for enhancement is a sense of 'body dysmorphic disorder.' This seems to be accentuated by social media. Social media may increase dissatisfaction with appearance and promote dysmorphia and the desire to use artificial means of enhancement. Use of silicone, Gore-Tex, polyethylene, titanium, polyurethane foam, polylactic acid, bio-protein glue and bioglass materials, for example, add to the body. This may be by injection or prosthesis. 3D printing will play an important part in the future. One can expect the prosthetic element will increase and expand as technology increases supply potential and demand. With ingenuity of marketing and possibilities of changing appearance and shape, especially if there are useful, functional, potential adaptations to a new digital environment, technosphere or technium then transhumanism emerges easily therefrom. There is seldom a social debate about such issues. It is unlikely that 'butt-lifts' received much prior discussion before becoming established. The sequence is probably more perception of a potential market, creation of product or service and supply. Similarly, we can expect supply first and questions later. Arguments about morphological freedom can be asserted on the basis of human rights and freedom of expression. ER also refers to Emergency Room. It is oddly often there or in surgery enlightenment may suddenly happen. I have listened to Pim van Lommel in conferences and read his work on NDEs. He knew his patients would clinically die during their heart operations and he decided to study them. Another greatly neglected writer is Benjamin Blood (1832-1919) who wrote *The Anaesthetic Revolution and the Gist of Philosophy* (1874). Surgeries are sites of alteration mechanical and mystical.

'Lagom' is a Swedish concept suggesting just enough. It suggests a sensible sufficiency. In some senses it is like the Goldilocks' 'just right' idea. Society suggests you can have everything you want, dream of and much more. Much misery comes from greed. Much suffering comes from attachments. Much pain comes from sin. How do we oppose this fantasy life that disenchants? Martin Luther King addresses this in his personalist way. He recognised the problem of facing life in a technical modern world. He argues for spirituality or religion and benefits that accrue. He stressed the need for a degree of acceptance of duality of light and dark and that the pendulum swings between both. He focused on self-acceptance of one's person and 'even one's looks.' Acceptance is associated with an inner peace. He says that we must sublimate rather than repress our emotions and sense of guilt through forgiveness of ourselves and progress unburdened. This would represent confession or repentance in his terms but MLK points to a general psychological principle relevant to the person. Likewise MLK said that he was not putting his faith in any small things but giving his allegiance to the highest force. Personalism allows for personal transcendence. Phenomenology was a method that sought to focus on experience as part of the story and escape from other limiting ideas. Personalism married idealism in some forms. It indicates that humans are exceptional and this goes against the idea of posthumanism. It means that we are somebody and not something. We interact with others as persons. MLK was informed by Bostonian personalism of Borden Parker Bowne (1847-1910). On the west coast George Holmes Howison (1834-1916) was a great, philosophical advocate of personalism.

The three point seat belt was developed in Sweden and spread by Volvo without reaping rewards they could have. We are familiar with the images of crash test dummies with seat belts hitting a wall when a vehicle accelerates. The Promethean, Faustian, Transhumanist 'Proactionary' force is a reckless, accelerationist one. Unfortunately, we are all in the same vehicle. The scientific drivers might be whooping with excitement and thrill of the ride but we are dummies in the test. Unlike artificial ones, we will experience impact. Scientific curiosity still gets its hit. A wall is a limit. Some limits are very hard when you hit them at speed. The apparatus or vehicle that involves insulation from reality may distort one's judgment about harsh consequences coming from what are essentially the inevitable results from pre-arranged forces. Many promises suggested by transhumanists are of dubious benefit. Why, for example, do we want someone to have a baby aged 75? What logic is behind these re-engineering strategies? When we cannot even explain consciousness it seems strange to distort the physical world just because we can, without a proper sense of relationship with its corporeal form. Tricks work because we are lured into traps through distraction and promises that appeal to our base emotions. We stretch bounds for no good reason.

When one sees ends to which techno-futurists will go and what they are prepared to sacrifice for their wishes, we realise that the tradition and reality of spiritual life not only needs resurrection but requires an embrace of a wider conception. Such a wider conception must allow a space above previous compartments. Such a space of spiritual exploration will be less defensive of perceptive insight. If we are walking into a death chamber wherein the human race is fundamentally altered and reduced we would do well to re-examine those who had spiritual insight. Thinkers like Rudolf Steiner (1861-1925) could be examined. His spiritual analysis of technology in terms of an 'Ahrimanic,' demonic force in his lectures and books seems consonant with the unfolding that may be explained in more rational and empirical terms. If one finds it difficult to accept literally that which he claims, it is still possible to relate his work to observable phenomena explicable in other forms. He criticised denial of the Christ aspect of Jesus. For him there was spiritual science which could be compared to ideas of Jung. He believed that dedicated spiritual study could combat Ahrimanic and Luciferian forces. One can adopt such an esoteric approach but even a robust, pragmatic attitude predicting consequences of certain phenomena provides a persuasive case that does not require resort to deep, spiritual insight. We have allowed our nature become subservient to technique and machines and accommodated technology so it assumes authority from access. Having yielded control and undervalued our nature and nature itself, we are vulnerable. The techno-scientific approach will make the technosphere conform to its control. Implications of complex contexts by simplification of vital forces through reduction are profound. Many thinkers realised this and reflect dangers of a technocratic tyranny erected on the firm foundation of our fear. The fear of missing out on a false, materialist paradise and the cultivation of impossible comparatives by corporatist propaganda bedevils us. When we get greedy we get vulnerable to predators.

Kafkaesque

"Someone must have slandered Josef K., for one morning, without having done anything truly wrong, he was arrested."
Franz Kafka
The Trial (1925)

Sometimes writers may mirror a force in society and sometimes they may predict it. De Sade represented the Enlightenment to a degree that is underestimated. He was a champion of 'reason.' Christopher Booker (1937-2019) identified de Sade as the start of a stream of sex and violence in literature that sprung from that time and ideas. It reflects also an attack on the psyche which has been ignored. De Sade creates a deeper feeling than mere disenchantment. There is sadistic sense of power provided by implements. This psychic disintegration reflected in *Justine* (1791) mirrors the effort to throw out the psyche and spirit. In many senses Kafka is bleak but moreso because of the difficulty of interpreting the intent. For me the work is reflective and prescient. A strange love of the machine and belief in its exalting powers even if executing someone can be seen in *'In the Penal Colony'* (1919). Kafka understood the deadly, absurd combination of machines, machine-mind-modes and administration in tyrannical offices and bureaucracy.

There is a startling relevance in the work of Franz Kafka to our present predicament. I recall people laughing at *The Metamorphosis* (1915) but I never found it funny. I did not like *The Trial* (1925) but found it strangely true.

The truth is that Kafka had a more accurate sense of the world as it would be than many others. He sensed the power and pointlessness that bureaucratic systems operating as a machine may create. His character in *The Trial* realises that the individual will be sacrificed to this relentless organisation. In *Metamorphosis,* the individual wakes up as an insect or monstrous vermin. It involved alienation, anxiety and isolation. Kafka develops a sense of a confined individual unable to act socially, forced to stay at home and made unwelcome there by other un-metamorphosed family members. The idea of an insect with extensions and protuberances but lacking the former aspects of the human body seems to anticipate the idea of a confined individual dependent on technology perhaps transitioning into a non-human or post-human but not through an exercise of free will.

The idea of transformation in form by magic without free will is a long and recurrent one. In Irish legend, the Children of Lir are transformed into swans for hundreds of years. In *The Tempest*, Ariel was imprisoned in a tree. There is a long tradition of the use of magic to trap beings in material. It even surfaces in the theology of the gnostics whereby in some accounts there was a trick to imprison spirit in flesh. There is a long tradition of spells cast and mythology involving the idea of a sinister magical transformation or imprisonment of one being by another. Supporters of transhumanism might therefore suggest that Kafka identifies discontent with the body and that the fear of transhumanism is merely an old wives' tale, fear or a superstition from a supernatural past that science seeks to leave behind. Opponents might just point to an ancient sense of the possibilities in evil magic.

The Trial involves a labyrinthine administrative system affecting the protagonist. The projection of a fog or mist or a system that disorientates and imprisons is a classic legendary and mythic power of the wizard. Some writers and artists have accurately foreseen what was going to happen. People presume that certain artists had a crystal ball. To some extent that is true. Certain writers are capable of intellectually predicting an accurate course of development. Others intuit or indicate, some psychic or prophetic. But other writers of fiction or non-fiction observe the force moving around them and interpret it. C.S. Lewis is one who noticed the milieu of scientists who sought to rule the world. He realised how serious and deadly was the intent of real people around him. That informed his novel *That Hideous Strength* (1945) and *The Abolition of Man* (1943). Likewise Orwell (1903-1950) moved in an environment where he had close relations with players in State authorities. To some extent therefore these people witnessed what was forming around them and where it would go. Some people think that certain streams in science fiction act as primers or precursors for the propaganda model to prepare us for what the industrial and commercial providers and producers intend. This possibility is arguably corroborated by media studies of people like Chomsky and others. With convergence of media, we can expect that the power-brokers in the era of hi-tech networks will seek to use power derived from their new positions to make sure that the mentality they have found success with will persist. Therefore we can expect that worship of technology or hi-tech power combined with means to manage others through cybernetics will result in an increased tendency to transhumanism.

Centralisation of power will be reflected in the system of technological intervention in the human. It will quickly become a matter of compulsion and not choice. We face the likelihood of compulsory intervention on a mass-scale to transform humans through inserting low-level implants while masquerading as public rationale for some common purpose of good, chosen undemocratically by groups of technocrats. The nature of hi-tech networks and their assumption of governance through acquisition of leverage in transnational spaces beyond regulation will allow imposition of transformative technologies without our consent, perhaps motivated by good reasons in the mind of the governors. The problem is that the narrow, materialist, networked, internetted, global village of a new industrial age will continue to create a world of a technical nature. In this world, the person is a lesser, messy force to be managed, integrated into a machine and turned into a machine. Assumptions that promote this worldview must be tested and rejected. Artists and writers who anticipated dystopia are more in line with nightmarish implications of authoritarian science as indicated in scientist's own words. Much imagination of the future was inspired by perception of the phenomenon of scientific players with such dark dreams. Hobbes suggested the *Leviathan* (1651) as a concept of shared sovereignty for the individual's protection consistent with ideas of the paternal state or commonwealth. The Leviathan we face is one emerging from hi-tech networks and techniques for a few. The Cyberspace Commonwealth instead comes by commercial conquest without consent. Its origin now makes few claims to legitimacy beyond opportunity. It uses confusion as a *modus operandi*. We are increasingly told we may be eating insects. We must also anticipate our metamorphosis into some hi-tech insect fixed in a chrysalis. The word 'origin' is suggestive of gold and metal. In the spiritual world there is an idea of the imaginal and of the emergence of the spiritual being from a chrysalis. This idea is adapted in Bernal and Aldous Huxley (1894-1963) in a mechanical sense in the idea of ectogenesis.

There is always a mechanical and a vitalist or spiritual parallel. It seems that science and the Enlightenment needed to replicate what was there, translate and then appropriate from spirituality, monopolise concepts and ensure that their institution could become the sole creative force in the universe as it displaced notions of other first causes. The Enlightenment siphoned from spiritual enlightenment just as they took ideas of illumination and illustration. Conquest of mortality promises a type of eternal life. Circumvention of natural birth and genetic engineering promises a new power of creation. Hybrids for diverse purposes have been created. Morphological freedom promises a Star Trek diversity of characters with little consideration of the implications. Before we know who we are however, we want to permit ourselves become something lesser than who we are. The dream of physical immortality is a pale and false promise that seeks to present an alternative to eternity whilst ridiculing the notion that our consciousness persists. It is driven by fear. It will be promoted by many who are either unaware of scientific or Enlightenment-constructed commitment to transhumanism or dishonest in their presentation of the reality. Morphological freedom is a recipe for a generally unpredictable dystopia of chimeras and hybrids which allows mass morphological imprisonment masquerading as its opposite. Licence is not liberty. We confuse the two. Licence of the few will be at the expense of liberty of the mass. The danger of being transformed into an ugly mechanical insect is greater than the possibility of transformation into a beautiful mechanical butterfly. The idea that something is Kafkaesque echoes dystopian adjectives like from the novel *1984*. It was part of a general realisation after World War 1 and the Russian Revolution that the growing bureaucracy that surrounded people and managed their world with machines, was not consistent with the individual's liberty. In the novel *We*, written at the start of the 1920's, by Zamyatin, we see the individual totally managed by collective rationality in glass domes and behind a green wall. The tendency of such collectivist, mechanical societies is to excise instinct, exclude nature and the organic in favour of technology, management and mechanisation promoted as progressive health and safety for medical reasons. Transhumanism fits into this trend as it comes from the nature of technique and technology and ultimately from a desire for control and governance consistent with cybernetics. It manifests in managerialism by elites with an associated cult of experts and technical interventions without genuine public or community support. An odd paradox appears to be that the pursuit of rationality leads to the irrational and paranoia that gives rise to the totalitarian system. In all this however, the force of automation and technique does not require a great conspiracy but the momentum of efficiency. As Vonnegut (1922-2007) indicated in *Player Piano* (1952) there is a serious loss of meaning to be considered in the search for material comfort in a machine-run world.

In relation to Kafka who perceived what was happening along with other writers like Lewis and Tolkien, it is important to note their concern with the demystification, demagicalisation and disenchantment of society. Such people were opponents of the insidious insinuation of the ideology of scientism that was emerging into the corporate hi-tech control of media. The messages delivered thereby must support the production system that makes the propaganda. Some thinkers realised that another contemporary mythology or fairy-tale system was needed. It was needed to indicate an alternative sense of reality and illusion. It is needed now in particular because centralisation and coordination of the public imagination made possible by pervasive technology becomes engaged with magic. Myth and fairy-tale in their manifestation as tools for the subconscious, collective unconscious, imagination and imaginal worlds are being substituted with replacements adapted to the system. Such magic as was embedded in folk tales and shared as communal experiences in enriching and educational ways, is mined, extracted and used to exploit the receiver by assimilating it into a narrower ideological structure. The magic that lay like veins of gold in the public imagination is re-shaped with some psychological recognisability retained to serve the interests of the private promoters and producers and restrain individual engagement with traditional narratives. By appropriating the inherent magical truths that pertain often to the collective unconscious or subconscious, a new magic is made that does not have the value of the whole context it previously may have had when integrated in a complex social system. Kafka and Lewis sought to write modern fairy-tales for adults and our future.

Snow White and the Seven Dwarfs (1937) is an example of a film that arises in discourse in relation to Nazi propaganda. Association between such extreme administrations, tyrants or totalitarian governments and destructive magic is controversial. Books like *The Morning of the Magicians* (1960) by Pauwels and Bergier became popular with particular groups. Much has been written about the occult interests of the Nazis. The Faustian connection seems clear enough however. When there is a rejection of any notions of morality in favour of an ideology, fantasy or fable that promotes a particular, perspective we can expect an enabling paradigm to be projected. Such a story or narrative will often relate to some causative idea such as the superhuman. The story may be for the elite controlling apparatus, priesthood or populace in general or a combination. A corporate apparatus requires a story or mission statement that acts as a cohesive agent. The appeal to the something which seems above or beyond or super, is related to the need to acknowledge that dimension especially as a justification for going beyond morals and tradition. Transhumanism deals with this idea of the superhuman. Writers like Kafka sensed a monster lurking in the system and pretensions that create it. He intuited the likelihood that a systematic, bureaucratic apparatus would end up with imprisonment, death and transformation of an ugly, mechanical sort. This is a warped magic whose power operates partly from re-creating illusions based on the archetypes dwelling in the darkness of our collective unconscious and subconscious. Watch and see how flying saucers or flying sorcery comes into consciousness as our masters decide appropriate.

Androids Dreaming of Electric Sheep in a Crack in the Universe

"We need magic to be able to receive or invoke the messenger and the communication of the incomprehensible."

Carl Jung
The Red Book (2008)

One writer who sensed what was happening was Philip K. Dick. He was not part of the scientific apparatus. I think he was more of an inspired prophet though many would disagree and see him as crazy. He was believed to be paranoid because he thought that the intelligence services in the US were sabotaging him. His perception of a 'Black Iron Prison' was consistent for many with those paranoid tendencies. Certainly there are people who have delusions but it is interesting that scientists, like John C. Lilly, who worked in this environment, and certain artists who worked in that milieu, pointed to the same phenomena. This was a sense that we would become captured, that we were prisoners and that the future was a dystopian one of mechanisation. Dick's analysis is complex and not as deterministic as suggested sometimes. His ideas were adapted and manifested in the films *Blade Runner* (1982) and *Minority Report* (2002).

In my view Dick was a mystic. He spent much of his professional life trying to make sense of a mystical experience with a download of information. This profound experience is totally consistent with the history of mystical experiences. When one reads *The Exegesis of Philp K.*

Dick (2011), it is clear he was a significant thinker more in the way of a prophet. He believed that the Roman Empire had never gone away and was revealing its pervasive presence during the Nixon administration in the US. There is a kaleidoscope of explanations, often contradictory, which he explores as part of an attempt to reconcile profound experience with his perception of contemporary reality. He may be defined as gnostic in many senses and his knowledge is part of a long life of illuminationism that is found in the perennial wisdom in Persia, Ancient Greece, Christianity and in all Shamanic traditions if one examines it sufficiently. If one is very sceptical about his mystical experience, it is possible to just examine the implications of predictions or contours of his worldview or vision as expressed in his novels and stories. I am not skeptical and see him as a modern wounded prophet or mystic. Philip K. Dick was very aware of transhumanism and posthumanism as technically likely. He identified a trajectory where prolonging of life and re-engineering of mortality were commonplace. In his works the human and machine distinction was confusing even for conscious agent themselves. Space colonisation occurs as anticipated by scientists but there seems to be a pervading, existential ennui and angst projected onto the future. He implies that issues of personhood, identity and who we are have not been resolved in the brave new, but psychologically-limited, world. Presumptions of greatness promised by science never seem to materialise or fully deliver their promise. The materialist vision may materialise but that is all it does with a lack of vision, aesthetic enhancement of the environment or intellectual or spiritual evolution. The advent of robots, cyborgs and androids complicate the remains of human existence without compensating for the loss of autonomy. Philip K. Dick saw what was coming. He is not universally dystopian. His machines may be more human than humans at times. But the sense of a dull dream permeates his novels.

Dick points to a direction to counterbalance the transhumanist trend. There is no doubt that many people considered him imbalanced. But that is the fate of many mystics. He was an accidental mystic in some ways in that he pursued his studies after a significant mystical event which is not unusual. It has often happened that one mystical event out of the blue causes the deep mystic journey. It does not mean that the person was not a searcher or explorer beforehand. An ineffable mystical experience punctuates a person's perception of reality so much that they often spend the rest of their lives trying to understand and comprehend it. *The Exegesis of Philip K. Dick* is an incredible exploration of that experience. That something is 'ineffable' does not mean that it is inexpressible totally but does point to incomprehensibility behind it. When one seeks then to corroborate, compare or compress one's own experience, one is likely to embark on a journey of engagement across cultures consistent with the perennial philosophy. The path towards gnosis beckons. The idea that the individual has a channel, access, connection to a higher form of revelation or knowledge, whatever one calls it, is described and affirmed. Although it may be confusing to the individual, they are certain of its impact and reality. That the person who undergoes a mystical experience may appear mad to others is common. That they often produce their greatest work to achieve levels of understanding of consciousness is critical. People who experience a download of knowledge or revelation with original content and comprehensibility in some compartments, understand that there is some truth about it. It may be about the future. Indeed, anxiety about the future may activate an assault on the senses. Mystic experience does come into the person as a real force. It is not merely imagined or fantasised when levels of affect are so high and palpable impacts are perceived. Some scientists suggest that there may be some evolutionary tendency to mysticism that asserts itself at times of crisis. A more standard view would be of an intervention of higher force through persons who were in some way receptive. There is a general and a specific receptivity. The general receptivity occurs through some feature or inquiring disposition or feeling in the person. The specific receptivity occurs through activation or rather de-activation, which may explain the work of Benjamin Blood. But mystical states may be activated by a range of factors. Without some underlying or general receptivity, specific reception may be ignored or remembered but not turned into spiritual fruit. This is what the parable of the talents of Jesus is about and not investment otherwise. Kabbalah is about receptivity. Receptivity often comes about through de-activation when one allows, or through accident, permits the universe permeate the mind. The filter of the mind is removed, mesh moved, the veil lifted.

Philip K. Dick was a modern prophet. He had a classic mystical experience that opened his mind. He foresaw where trends manifested in California were going. He understood technology. Because of mystical experience or gnosis he readily perceived the implications of hi-tech developments. While his investigations are seen by some as crazy, speculative ramblings there are profound insights and awareness in his work. His life shows an interesting connection between spirituality or mysticism and technology. If spirituality or mysticism is the real noetic force it claims then it must assert itself. It is clear that more scientists are emerging who recognise the reality of mystical experience. Hopefully, re-assertion of spirituality may re-balance destructiveness of the materialist agenda. What the prophets or imaginative artists foretell is that these forces are real, active and transformative. They are happening now. While we may wish to merely see a packaged narrative in the propaganda machine, we should apply a more rigorous philosophy to test assumptions and presumptions behind mechanisation of the individual or person. We cannot ignore the inevitable impact of transhumanism and pretend that we are merely observers of benign forces happening organically around us. It might be that our society conjures simulacra and cyborgs because we have failed. Forces already unleashed are leading us in a direction of very limited options. That is why we must awaken and be aware that arguments to the effect that we are maladapted mentally to evolve will seem proved for their proponents. Writers like Dick witnessed, interpreted, intuited and imagined what would happen. He did so often on the basis of an assumption about the recurrent tendency to utilise technology without improving our spiritual foundation. Dick was led personally to the sources of theology and Gnosticism to try make sense of a future he sometimes fairly accurately anticipated. His work not only acts as prediction but in the classic tradition of prophecy it seeks to give meaning.

Eichmann's Children

"'The Machine,' they exclaimed, 'feeds us and clothes us and houses us; through it we speak to one another, through it we see one another, in it we have our being. The Machine is the friend of ideas and the enemy of superstition: the Machine is omnipotent, eternal; blessed is the Machine."

E.M. Forster
'The Machine Stops' (1909)

Eichmann in Jerusalem: A Report on the Banality of Evil (1963) by Hannah Arendt was remarkable for failure to find an obvious monster. Totalitarian machines require uncritical dullards who have a total technical focus and a tendency to concentrate unquestioningly on the quest to achieve some specific objective. Tyrants love technocrats unburdened by a sense of ethics or morals. An assumption that those who manage us must be constrained by higher values and only want our best interests is a naïve fantasy of an infantile nature in view of ancient as well as recent history. Many critics have pointed to rapid advance of a range of apparently psychopathic or narcissistic elitists and experts in a managerial class that chants scientism slogans and fits into an evolving apparatus of transnational governance beyond reach of effective regulation that interacts with military and security establishments.

Günther Anders is a neglected critic of technology. While Heidegger is a significant critic, Anders was one of his students and also of Husserl and he was associated with Adorno (1903-1963) and married to Hannah Arendt. From his experience in Germany, he was very able to examine the better theories and match them with reality. He saw flaws in Heidegger. Anders paralleled Arendt's study of totalitarianism. After Hiroshima, he realised that the shadow of that cloud would cover humanity for ever more. He developed his theories of technology further to accommodate the impact of tv. Some describe his approach to technology as 'post-Marxist.' C.S. Lewis would share much of the deeper analysis put forward by Anders. In *That Hideous Strength* (1945) Lewis identified progressive entryism or takeover in his fictional university which was related in some ways to the rise of the culture of scientism which he painted as sinister, demonic and conspiratorial. Anders was a humanist and a personalist.

Anders communicated with people involved in some way with these catastrophic events. He communicated openly with the son of Eichmann and one of the men responsible for dropping the bomb on Hiroshima. Like his ex-wife Hannah Arendt, he was very interested in Adolf Eichmann, who had been kidnapped in Argentina and flown to Israel for trial. The legal process and studies of Eichmann led to a sense of the banality of evil. Eichmann pointed to his obedience and lack of power over the elite which ruled the Nazi Party as an excuse or description of his sense of lack of culpability. There was a very strong sense of a bureaucrat doing a job. The sense of a machine and of being a cog was a real one because it described the mechanised clockwork universe, where people fit in. The person becomes a part of the megamachine perhaps with a steering puppet-master at the head of the party. Anders saw the danger that we can all contribute by our participation in this machine which must become totalitarian when based on a Promethean will and a fetish for technology. The idea of the fetish can be related to Marx, but it is not so unusual. It indicates this remarkable love of the machine and technical objects that drives social machines and often is destructive. People are part of the machine. We can all become Eichmanns.

The 1950's was a critical time when the war-machine grew, switched to consumer goods and mass media grew up. The date of increase of tv in the UK was associated with the coronation of Queen Elizabeth. This time was linked also with AI and transhumanism. There was something in the air. Cybernetics was casting the contours of webs and nets over the mind through the machines of humankind. Anders understood the dramatic influence of tv as mentioned. He also wrote about how the machine led to the end of mankind. In *The Obsolescence of Man*, he explained how there was an inevitable tendency to produce machines. The atom bomb changed us into Titans that were losing humanity. In his thinking there is a loss of humanity which may be the result of over-concentration on one particular aspect of the human. This trend towards technological power turned us into something else. In Arendt's work there is an implicit focus on the person. She looks at the persons that contribute to humanity and the persons of those who destroy. She recognises the individual as a moral agent. In the personal qualities of individuals, the best of humanity must manifest also. The philosophical or psychological base of personalism seems to help people better identify the power of impersonal forces or processes such as technology and bureaucracy.

The producers and puppet-masters of the show betray a nihilism and misanthropy that mystics like Blake warned were infused in the deadly, scientific approach. The much promoted rigor of the scientific method can become a rigor mortis for the human species. A degree of misanthropy and nihilism seems to permeate some scientists who condition the public. Undervaluation, devaluation, de-humanisation and de-personalisation of the individual drive the big, careering vehicle. Strange arguments from the fringes of the environmental movement suggest the desirability of the end of humanity for the sake of the environment. Peculiar deconstructions of posthumanism are posited on selected values that have no base other than being the logical extrapolation of an underlying, existing strategy. A type of humanophobia threatens to trump all existing and probable phobias. Eichmann or the Einsatzgruppen, Beria in the Soviet Union and so on, operated in systems. The scale of these systems is based on cultivating opposites and promoting destruction of people through de-humanisation. The cavalier treatment of the person in scientific discourse will have consequences in the mind of some humans and denial of our essential nature and importance of the individual is sowing dragon's teeth. Anders warned about the 'Promethean gap' wherein our ambition exceeds our ability to manage our creativity. Our grasp on the reins in the rodeo is loosened in slow motion. He warned us not to be cowards. There is a surprising amount of cowardice not least among scientists who champion their ability to think critically. Science will contribute constructs once again to policies that treat the human in a less than human fashion.

Science, even workable mathematical theory, can be part of the problem. A desire to govern the natural world and people is a fundamental aspect of the instrumental impact of materialist science. Commercial interests impact on outcomes and influence institutions. The scientific, commercial and managerial apparatus is calculated to exhaust the planet and our human nature. We will be ransacked to feed the Machine. While philosophers like Anders identified the phenomena, creators of instruments or technique also identified how their tools could be used in strategies akin to sorcery. The hi-tech infrastructure that assimilates will make us similar, to fit in, until we become simulacra. We are being taken in by machines, systems and techniques. Many sense that we are totally controlled by political, scientific and commercial machines run by mandarins, managers and merchants and this will be manifested in the form of more managing machinery. Worse, people sense the reality that many of our masters wish to turn us into machines because they see us thus already through an obsessive mentality of misanthropy. More and more we will witness a new scientocracy presented as inevitable by people who claim authority without demos, who are willing to employ the straitjacket of hi-tech networks to control us. Anders was mostly driven by William Stern (his father) and insights on the difference between persons and things which began about summer 1900. Stern's personalism anticipated answers to the challenges that we face. Anders saw that the problem was more acute. There could be no focus on the person if the apparatus, bureaucracy or system created by technology to manage us totally rendered the person obsolete.

The evil monster may manifest in the dull appearance of the bureaucrat as C.S. Lewis warned. Eichmann seems to have been a lot more conscious of his role rather than being a mere bureaucrat. As the architect of Holocaust he seems to have regarded the whole operation as a giant death machine. The persons exterminated were of no moment to him. A machine attitude and analysis of systems as mass biological problems to be managed for particular purposes is a product of a mechanistic mind-set. He argued that his work was merely a medical operation at times. This ideological racism also grew out of the ideas of scientific racism and eugenics that appeared in the heart of the British Empire in places like the Eugenics Record Office in Gower Street, near Russell Square set up by Galton (1822-1911) in 1904. This was a smaller operation compared to the Eugenics Record Office in Cold Spring Harbour in the US. Eugenics seeks to use data to design humans based on genetic characteristics. Genetic-editing has become a significant technology in recent times. The experience of eugenics was related to campaigns to abolish traits and people deemed undesirable and reinforce ones perceived desirable. I do not suggest that individual transhumanists are inheritors of this meme. Nevertheless the search for superintelligence usually relates to the genetic context. Transhumanism seems very close to the ideas of eugenics in some ways. While there may be legitimate interventions in appropriate cases, the idea that widespread genetic manipulation could be used as an instrument of policy involving compulsion as part of an ideology of scientism is a very dangerous one now.

The US Eugenics Record Office was linked to the Station for Experimental Evolution in the same place. This shows the close connection between techniques of controlling and intervening in biology through data and the idea of improving living beings and people so they evolve. As wealthy landowners in the British Empire had bred the upper class, horses, dogs and plants to improve them for particular purposes, the idea of artificial selection instead of natural selection was one promoted by science. A question arises as to how a group of people can decide what is good for humanity and who should live or die and why? This resort to playing God with a degree of certainty about the rightness of their decisions is a strong tendency in people who gain technical power through possession of certain knowledge, ideas or instruments. It was second nature to the imperial elite in London and other capitals. It seems like magic. The genetic industries continue to grow and continue to refine technique. Eugenics thinking informed the Nazis. We are now supposed to assume that people who claim some right to create superhumans and who want to do so consistent with a positive eugenic mode while talking about evolution are nothing to be concerned about. I have a degree of understanding of people who choose to do things to their own bodies and of course celebrate restorative prosthesis. Many of these other experiments in control are performed on animals and will also be on humans. One of the main proponents of transhumanism nearly a century ago made clear that docile humans could be experimented upon without their knowledge. No need to worry about transhumanism at all. As scientists like James Lovelock corroborate that we are creating a 'Novacene' of hyperintelligence wherein we will be like plants to machines it is clear that our end is near and many have contributed to it unwittingly or not.

Humanity Rest in Pieces: The Fourth Industrial Revolution

"I'm talking about the real owners now... the real owners. The big wealthy business interests that control things and make all the important decisions. Forget the politicians. The politicians are put there to give you the idea that you have freedom of choice. You don't. You have no choice. You have owners. They own you. They own everything."

George Carlin

We have crossed the threshold into a human endgame. It is clear there is a strong, centripetal force towards global governance. Global governance from hi-tech industries and technique are fundamentally related in a circular way. Hi-tech drives global governance movements and provides means to implement it while emerging from the military-industrial complex. The World Economic Forum at Davos, Bilderberg, Bohemian Grove and a range of UN bodies, the Trilateral Commission, Council of Foreign Relations, Billionaires and big business and others create the core construct or constrict of emergent global governance networks. Hi-tech, security companies and Big Pharma in a symbiotic relationship can expedite construction of technical infrastructure of a global governance system of us as objects or things. Surveillance and control allow management without oversight, responsibility or rights. Unfortunately, this is just the start.

Klaus Schwab of the World Economic Forum has written about the 'Fourth Industrial Revolution.' This indicates the future of the technological environment we will live in. It is clear when one follows such arguments that there seems to be a sense of *fait accompli*. It is not that these people are predicting what will happen but it seems that they are advertising policies which have been agreed. Edward Snowden talks about the biggest conspiracies taking place in plain sight. This is entirely consistent with the agenda of H.G. Wells as indicated in *The Open Conspiracy*. This is echoed in J.D. Bernal's *The World, the Flesh and the Devil: An Enquiry into the Three Enemies of the Rational Soul*. Therein, this idea of the triumph of science, through the secret rise of scientific corporations was indicated. The materialist paradigm and mechanistic mind-set that transformed into an ideology and scheme for global governance has now transmuted into a policy that is being implemented under the guise of an inevitable unfolding of technological and industrial development.

Vague indications of supposedly ineluctable events suggest an inexorable process promoted by big business above Governments and in control of regulation. Principles of quantification and surveillance will be projected onto the individual without qualification and without privacy concerns. The tendency to 'robotize' humanity and deprive us of our heart and soul is acknowledged. The implication is that the best we can gain is through glad acquiescence to the pervasiveness of technology and technological control of us. The sense that we have no choice is very real. There is no pretension to prevent this massive loss of sovereignty. The message is clear. You will be integrated into this system. Schwab co-wrote *Shaping the Fourth Industrial Revolution* in 2018.

It is clear that the process of hi-tech convergence through the 'internet of things' integrates us into the great machine until we become machines. Those who are dedicated to novelty, machines, control, order and governance will concentrate their resources and combine science, data and mathematics with the process and philosophies of deconstruction that produce a scorched earth on which a new world can be built. It is not new.

> *"In looking for a way out of the present extremely wasteful merchandising traffic, and in working out organization tables for an equitable and efficient distribution of goods to consumers, the experts in the case will, it is believed, be greatly helped out by detailed information on such existing organizations as, e.g., the distributing system of the Chicago Packers, the chain stores, and the mail-order houses."*
>
> Thorstein Veblen,
> *The Engineer and the Price System* (1924)

If it was merely a case of sensible, restrained people who really want the best for us, suggesting a reasonable way forward that might be adhered to because it is so persuasive, then such schemes could be taken seriously. However that is not the case. It is clear that certain technocratic and 'technetronic' strands seek to combine technique with machine networks to control us. Having controlled us, the Faustian fun can continue. We might think of the 'imp of the perverse' intervening. But better still, just look at what scientists have said about assuming control. The mad scientist is not merely a fictional concoction but represents a real force that runs us. Jack Parsons was a genius with rockets who performed significant black magic operations also. His role in the Jet Propulsion Laboratory and space race is well known. Other minds that had a big impression through NASA were Werner von Braun (1912-1977), the Nazi engineer. Morality is ignored if scientists serve or accord with your objectives. Parsons was an occultist-scientist. He was a great outsider and part of a triangle involving L. Ron Hubbard (1911-1986) and Aleister Crowley. Perhaps the dedication of Parsons to magic argues against the idea that the scientific establishment could represent a black magic approach. Parsons was a part of a 'Suicide Squad' and he died very early. For some he perhaps represented the Promethean, Faustian and even a Satanic, Luciferian approach. Parsons was a brilliant scientist in whom occultism and science went together. His occultism probably informed his innovation. He sounds like a genuine explorer. But it does point to that curious, reckless, bold archetype who often innovates in science. So instead of careful, considered scientists we have two passengers in the front perhaps high on rocket-fuel fumes with nitroglycerine in their pockets blasting Queen "There's no stopping me..." and babbling something about Babalon Working. Even Parsons was sidelined in his life. It is not so much explorers that determine things but experts who take over systems the innovators create. The governance of the world is increasingly determined with technique by the few and for the few.

As the subtext of transhumanism or the force behind the front becomes revealed and as the magical use of network wizardry to promote elite interests through mass networked assimilation of us, the misleadingly wide claims for transhumanism will be exposed to those who care to look. As well as the bodies on ice and cyborgs, they will claim or imply that all hi-tech medical advances and health maximisation technologies, ultra- or super intelligence, social justice and space-exploration are somehow transhumanist. Indeed, there is a suggestion that all the intellectually homeless, whose potential homes in eugenics and futurism were abandoned for obvious reasons, can find a new place in a movement that is defined by a general sense of love of technology, science and an interest in physical survival. Ironically, the more hard-sell and attempted conceptual monopolising they do, the more they seem to corroborate the real agenda of the creation of a scientocracy as proposed by scientists, particularly of a Marxist persuasion. In transhumanism, the possibility of a materialist half-way house for extreme left and right is made real by the simple will to power as concentrated on control of technology and technique and the application thereof to total governance of humanity, pending their separation, down-grading and demise. If you are against this you will be vilified as a fool, Luddite, free-rider, extremist, religious nut, crank, down-winger and so on. The 'up-wingers' will demonstrate that the argument that the technocratic, technetronic movement is about energy control and that this represents potential tech-totalitarianism was more than likely right. Apparent national power bases are now mere tentacles of the global, undemocratic monster of bio-tech scientism governance.

It seems that the work of certain theorists like Marcuse was prescient. Capitalist systems are revealing their fundamentally totalitarian nature. Marcuse in his work *One Dimensional Man: Studies in the Ideology of Advanced Industrial Society* (1964) perceived the totalitarianism inherent in advanced, democratic capitalist systems of soothed stupefaction whereby the workers were assumed into systems which appeared to offer freedom. These societies were only free in a limited sense and lived in an illusion of substantial options while really having limited choice and a huge, existential threat to enforce it. Opposition was not real. Now crony capitalists tell us we must tighten our belts as they enjoy excess profits. We must stay at home while they go to space. We must consume less while we have been channeled to so consume. It seems to have been a chimerical charade. Now that magic carpet may be pulled from under us. Marcuse identified technological rationality as a major, restrictive way of thinking that technology caused. However, such an approach arises equally in communist systems. Marcuse also explained how commodities for consumers become extensions of them. This is another sense of prosthesis and merger. The consumer identifies with the product and may become part of it in a way, as with the motor car. It seems that it does not matter whether it is left or right, the force of technology creates an objectification of the individual and alienation. Yet if we do not have a strong concept of personal or spiritual values or qualities there is little alternative. As the present US President remarks that he needs US troops home to create a 'new world' we can expect to see their expertise directed to enforcing the new status quo.

If you are sceptical about the existence and persistence of malign or manipulative forces exerting control over your world then merely examine the impact or effect of government intelligence services, pharmaceutical and tech-giants in the last century. Look at the horrendous 'sacrifice' of life to systems of finance and industrial production that are inherently wasteful and anti-human. Look at the deliberate and widespread deception. The security, intelligence, military, pharmaceutical complex have clearly been one which Presidents feared. Ideas of 'intelligence' and magic have always been linked. The idea that destructive magic is a marginal activity is inconsistent with the evidence of the link between magic, sorcery and power from an anthropological perspective. Apparent attacks on say witchcraft or alternative occultism by the State as say in the Protestant United Kingdom may arise precisely because the rulers believe in the power of magic just as King James personified. Intelligence, secrecy, circles, power, rituals, symbols, disinformation, control, potions or 'shock and awe' are common to governance systems and black magic. As well as the artists mentioned, some film-makers represented the forces we face in audio-visual form.

Thus to finish this section it is worth noting the explorations of Stanley Kubrick (1928-1999) who anticipated many of the issues we face today. Kubrick developed Arthur Schnitzler's (1862-1931) novella *Traumnovelle* (1926) into *Eyes Wide Shut* (1999). He suggested the unconscious of us may manifest in real world occult cults. In *Dr. Strangelove* (1964) he indicates how a psychopathic madness in quantitative management may de-personalise people. He understood how certain scientists could shift into a de-personalised mode when they abstract themselves and treat others as de-humanised abstractions. Kubrick also directed *2001: A Space Odyssey* (1968). This was based on the novel by futurist Arthur C. Clarke (1917-2008) who predicted human-machine merger. The film is seen as transhumanist by proponents. In keeping with that narrative, it suggested an alternative prehistoric path of evolution predicated on intervention by an alien, hi-tech force. This is consistent with the suggestive predominance of technology as a godlike figure and identification of the human essentially as a hi-tech assisted ape. The same hi-tech force improves the human again with another intervention after the poor human had lost a battle of wits with an intelligent machine (HAL) it created. The transhumanist tendency to reduce the human or divine to a level under hi-tech is evident. In *A Clockwork Orange* (1971) based on Anthony Burgess's (1917-1993) film of 1962, Kubrick explored the idea that bad qualities in humans could be erased by hi-tech programming. The suggestion by Burgess is that such editing would make us less than human. Transhumanists suggests a similar editing. We are now told that taking something away is an enhancement. Such additions seem to be instead an accumulation of force in favour of machines and minds that love them at the expense of the person. Such a desire for power may be consistent with pure individualism which glorifies the 'objectivism' of Ayn Rand. Such materialist individualism will sacrifice spirituality, solidarity, community and personhood to attain a selfish system of control for the good of the few.

THE FOURTH PART

Re-asserting the Real Qualities of People

The immeasurable of us is both the source of pleasure and the real treasure of our consciousness. An obsessive, opportunistic impulse to measure and quantify and ignore that which cannot be easily assessed is de-meaning and de-humanising us. Mesmerised by measurement and models we have lost sight of reality. This is a disastrous path previously evident in biological determinism that now has morphed into bio-tech determinism. This section seeks to indicate some ways that we can begin to challenge or nudge concepts which underpin ineluctable growth of the technosphere and Empire of Scientism. It suggests caution about over-reliance on quantification to the exclusion of qualitative factors. It re-emphasises the need for respect for human, individual personhood. It re-iterates my call for a concept of the 'animasphere' as a superstructure in order to facilitate harmonious interaction consistent with perennial philosophy. The immeasurable qualities and ordering of the human, personal potential in a meaningful, cosmic way need to be asserted in the face of a deadly disease of obsessive quantitative disorder and techbondage. We are now entering a trap of excessive quantification as part of an apparatus of control if we do not assert sovereignty of our inherent qualities.

Quant T.R.A.P to Quantum Q.U.A.L.I.A

> *"If an increasing number of people become fully aware of the threat the technological world poses to man's personal and spiritual life, and if they determine to assert their freedom by upsetting the course of this evolution, my forecast will be invalidated."*
> Jacques Ellul
> *The Technological Society* (1964)

Being able to apply calculations for sensible and necessary purposes is of critical significance in society. Nevertheless, constant, recurrent, systemic failure of figures, calculations, statistics, polls and predictions is also evident. Appearance of authority through the cloak of calculations can replace sound judgment. There is some connection between the worst and most inhuman aspects of science and excessive belief in comprehensiveness of calculation. Galton loved numbers and promoted eugenics. As Gould (1941-2002) makes clear in *The Mismeasure of Man* (1981), apparently sound quantitative methods may merely promote our prior prejudices and seem to make authoritative that which was undeserving of it. There seems to be some psychological connection between the mind that requires control and predictability and believes intently in the reassuring presence and apparent truth of numbers, theorems, computations and formulae to reduce uncertainty even if such symbols are merely marshalled in a Procrustean re-alignment to fit people into predictable patterns.

In the UK, there was a lot of work that prefigured that of Wiener on cybernetics in the US. Cybernetics is about steering or governance mechanisms looking at animal and human behaviour and applying principles of feedback and computations. After the Second World War, the Ratio Club was founded with a diverse range of academics and professionals to study similar ideas. It is important to realise the more inherent quantitative bias in the word 'ratio' in this context. For a nation on rations during the war, it was not an unusual word. Clearly it is related to reasoning and rationality. This 'ratio' was what William Blake had always criticised. From correspondence, before the Club was established, there was a reference to a 'Machina Ratiocinatrix' or the calculating machine after Wiener. They were appropriating the computational and statistical basis of reasoning. This is usually juxtaposed against an older form of divinely-inspired reasoning, revelation or the imagination. Some see this as a function of the left-brain, right-brain issue. The computational dimension of thinking or reasoning is a very narrow one. It contains selection biases and focuses on stimulus and behaviourist tendencies. When people talk of the rational as being consistent with quantification and computation primarily, we must be aware that much may be left out. The 'quantified self' is the emergent person in the digital universe. Science becomes a mere calculating machine.

The emergent Empire of Scientism as a techno-totalitarian system of governance is partly caused by a utilitarian, instrumental focus in the worldview and group-think of a corps of positivists, ultra-materialists, managers, makers and manufacturers of things combined with consistent messages projected by the associated global governance system's propaganda model using the tyranny of metrics and quantification. There is a vicious cycle of quant-biased management that has increasingly informed our governance. Those who correctly criticise quantitative method may fail to articulate the possibility that the misuse of management tools can be deliberate sometimes as part of an ideological strategy. Scientism may be a cover for deliberate disestablishment. For example, the idea of 'welfare capitalism' espoused by the former Trotskyist Richard Rorty (1931-2007) in his essay on 'Trotsky and the Wild Orchids' (1992) was an ideological tactic calculated to lead to capitalism's demise. The appeal to data, statistics, figures, polls and surveys as a sound basis for complex public policy choices can be captured and compromised by political actors who use the aura of authority of the quantitative to achieve results. The people that call for transparency may not be transparent, for rating may not be rated, for accountability may not be accountable, for performance may not be performing. Consistent with work such as *The Tyranny of Metrics* by Jerry Z. Muller (2018) I suggest we move from the quant TRAP to QUALIA to start recalibrating our thinking.

Transhumanism is one manifestation of the scientism ideology and technique and bearing them in mind and reflecting on a better paradigm may break the cycle or the worst elements thereof. If I was to identify a prime, private lever towards transhumanism it would be the fear of death. Furthermore, the fear of death is associated with a lack of belief in the persistence of spirit. Transhumanism is an attempted substitution of the predictable mechanical or machine for the possible mystical. Transhumanism seeks physical immortality because of a presumption or belief in mortality. Transhumanism seeks transformation of the body out of a failure to accept oneself. Martin Luther King identified self-acceptance as a primary consequence of religious or spiritual belief.

Undoubted success of the scientific method makes people make a mistake of extrapolation. The fact that quantitative information and measurement is useful, particularly in the natural sciences, does not mean that it is always useful. That which is measured may not be the most important thing. It may be measured because it is measurable. Measurers may become more important than that which they measure. People may play games with measurements and muck up the system. Operation may be compromised by measurement which is expensive and has a chilling effect. Tacit and practical knowledge or know-how may be lost or under-valued. If you are regarded as a mere quantification or calculation you may be regarded as an incidental externality. Dangers of quant-bias in the hands of bad actors or shallow thinkers combined with control via networks and the ideology of scientism create conditions for networked transhumanism. So I suggest a shift of awareness from a TRAP to a QUALIA approach.

TRAP to T.R.A.P

Transparency
Rating
Accountability
Performance

All these sound good. However they may be compromised in their use, so one must be careful in assessing them.

(a)   Transparency

We hear about transparency and assume it is good. There are certainly benefits. However critics point out that some things are best left alone. Jung prized interior space as well as exploration. Interference may impair them, just as we do not want light in a photographic dark room. The idea that everything should be visible, measurable and examined to be better is an argument of surveillance capitalism and scientific socialism. At the same time, there is a lack of transparency about covert global governance. Most of the apparently significant incubators of change at a global level are informal and secret or confidential. Key players in internet management meet. Scientists often meet to create new movements such as in relation to cybernetics. Decisions are increasingly beyond democratic scrutiny. Algorithms are now programmed privately to manage us. Transparency is a conceptual tool whose selective employment must be watched. Our lives are recorded in a surveillance society which takes our privacy while the functions recording them are done secretly. It is a two way mirror. 'Neuromania' is also increased by technology that only appears to identify consciousness. Thus we have paradoxical, selective and inappropriate transparency.

(b)   Rating

The word 'rate' seems to have some etymological and mathematical relationship with the idea of ratio. We see ratings everywhere. We are lured into rating others under the illusion that we exert power. We even pay for a proxy power to play ostensible judges through phone-ins, for example. Ratings can be easily manipulated or skewed. There have been spectacular failures in predicting elections and financial ratings. Ratings used to be more important for selling programmes or products. Ratings can be skewed by irrelevant measurement system. Associated with rating is the creation of models. Models may distort if they do not rely on robust assumptions and if that which is measured is not relevant. The appearance and aura of authority may make ratings mislead such as with IQ measurement. Similarly, psychologists such as Albert Ellis (1913-2007) explained the need to not rate yourself but look at your deeds and actions. Some things are finite and measurable and some are not. Nevertheless everything will be 'upgraded' including you. Social Credit Scores are on the way. Don't overrate rating.

(c)   Accountability

Counting is not accountability. That we can count things or objects or quantify or measure is an important part of accounting. But the notion of accountability implies calling people, especially those in power to account. Countability is not accountability. Those that count and measure may not be accountable. The recent financial crises saw many institutions fail despite employing wide massive countability, mathematics and quantitative ability. Yet few were held accountable. The individual person without power is often held accountable through some countable process while those employed as counters professionally escape accountability. Those things that are most valuable are not countable. Such processes may allow damage be done by making numbers more important than the quality of responsibility or the system itself. The global financial system is subject to manipulation because there is no accountability and abstraction, complexification and symbols obscure.

(d) Performance

Performance indicators are another possible trap. They may set up false goals, misrepresent what is important and fail to identify what is. We might have a high-performance car that destroys the environment. We may ask whether some bureaucrat is performing well by having futile indicators while significant activities are ignored. We seek to grow and evolve, look at what we do and how we act but must be careful not to focus on measuring performance of things which are not important. Economic growth targets can be wasteful or destructive. Public medical measurements became suspect in recent times.

So there is a trap in this T.R.A.P. Such traps are part of a bait method of persuasion. Such persuasion may be a type of unintentional diversion. They may also be akin to games, magical manoeuvres, mantras and incantations to achieve a result. Politicians and power-seekers may set up targets, tools or techniques to conjure an image of reality. The concentration on those things may distort reality. Repetition may hypnotise and mesmerise. The apparatus and aura of scientific authority may impress and press conformity as scientists like Milgram (1933-1984) and

Asch (1907-1996) have studied. The technological society is driven by the software of technique as much as the hardware it produces and applies. If there is a strong ideological force behind transhumanism, then techniques available to it in a networked society are legion. If there is a move to networked transhumanism so that we will be transformed to adapt to the network, the power exerted on us will be overwhelming. Our controllers have AI networks that we are increasingly dependent on. We are dependent on mobile phones. We are being integrated into a global bio-medical apparatus of information that will be very intrusive. New technology will make it easy to link us through implants without consent. If we are aware of the trap of quant-mis-information and manipulation then we could also be more conscious of the opportunity and need to focus on qualia. I use another acronym.

QUALIA to Q.U.A.L.I.A

Quality
Universal
Affect
Light
Intelligence
Actualised

Qualities are aspects that may escape quantifiability. John Locke, Descartes and others discussed the nature of qualities. The idea of qualia is more specific and can be traced as a term to C.S. Pierce (1839-1914) who was a pragmatist. They refer to the subjective experience of something consistent with a phenomenological approach. The term is older in use and sometimes meant the quality of perceiving qualities. Qualia relate to experiences of the world. They may be ineffable and intrinsic. In that sense they share something with the phenomenon of mysticism. Dennet who wrote 'Quining Qualia' (1988) and has the daft position of denying consciousness seems to seek to deconstruct qualia also. This seems part of a cult of scientism calculated to deny personhood. This is the needlepoint of the attack on personhood. A certain thrust in science is hellbent on deconstructing the person and concepts that define them. This reductionist madness is a calculated policy. Once you have denied certain qualities or qualia, you can deny experiences and ultimately existence. This shifts the burden of proof to facilitate disregard of our humanity. Physicalist critiques suggest a philosophical zombie argument to distinguish personhood from a simulacrum without experience. Transhumanism threatens the mass with philosophical zombification because it does not acknowledge qualia. That is why in rejecting a quant-trap we should embrace qualia or qualities. Personhood is real and philosophical denial or use of quantification will not displace qualia. The Procrustean solution is certainly about adaptation. You are forced to adapt to the bed you must lie on. We should begin with robust protection of the person and accept qualia while resisting sophisticated pseudo-science masquerading as philosophy. In my view, qualia are the interface between the external world and the consciousness that flows within. They are the end of the outside world in a way and the beginning of inner consciousness. Qualia are pinholes to admit light inside a camera obscura and from within to see a picture, inverted.

Hence Q.U.A.L.I.A.

(a)   Quality

Qualities may be unquantifiable. The highest elements of human nature are not easy to quantify. Personhood and qualities of creativity and imagination are unique and powerful but difficult to predict. Essential qualities of personhood of our highest awareness are beyond quantification. Consciousness is a fundamental *a priori* presumption that has priority to quantification. Analysis of experience is not the thing itself. Reflection of experience in accoutrements of science should not be confused with consciousness upon which sensory impressions are made.

(b)   Universal

Perennial philosophy deals with truth universally acknowledged across culture. Most religions and spiritual traditions have some basic ideas of reciprocity and humility or recognition of supreme consciousness. Nearly all mystical and mystery traditions indicate that the universal is contained in the personal. Posthumanism and transhumanism often attack universals.

(c)   Affect

The heart and feelings are an important part of the whole construct of the person. They do not triumph over the other elements through a soft, fuzzy individualism but a deeper recognition of one's own consciousness and others. Affect refers to some irreducible quality of feeling that is part of consciousness, not purely physiology. Love matters.

(d)   Light

The universal, spiritual experience of light was the source of the idea appropriated by the Enlightenment. The light of reason and illumination is a version of the deeper sense and spiritual experience of light. Light is still a mysterious subject that has driven science. We need reason and spirit working consciously together.

(e)   Intelligence

Intelligence may be indicated and quantified in some ways. Nevertheless, as argued by Stephen J. Gould in *Mismeasure of Men,* humans can be mismeasured. Intelligence was part of intellect which included a sense of the person before it was whittled down to computation. Scientists create tests to identify more people like themselves. Re-conceptualisation of intellect as involving spiritual consciousness will oppose a mismeasurement and misrepresentation of intelligence. Ratiocination and calculation should not be the sole criteria of intelligence.

(f)   Actualised

Maslow developed the idea of the self-actualised person in *Motivation and Personality* (1954). His idea is linked to Kurt Goldstein. Personality involves and stresses the significance of the person. It is difficult to contemplate the person without some sense of whole significance thereof including a mystical or spiritual dimension or at least a profound philosophical or psychic one. The consequence of a lack of personal focus is a de-personalising and de-humanising force.

Hoffman's theory of conscious agents is based on computers consistent with a classic materialist perspective and equiparates computation with consciousness. A better attempt is Federico Faggin. He has awareness of the significance of mystical experience which informs his worldview. He is a renowned inventor who developed the first commercial microprocessor. Faggin is developing a theory of conscious agents which is based on quantum entanglement. His theories seem based on sound groundbreaking, theoretical science with solid first-person, subjective experience. As a result, he is focused on qualia as critical. The synthesis of knowledge of quantum physics and quantum information allied to notions of meaning and semantics put forward by him may unify qualia and quantum theory in a comprehensive theory that moves beyond materialism and respects the person. If emerging theories, such as this, are correct, inevitability of the Singularity and digital uploading may be rendered redundant. Materialist science is undergoing a change. Post-materialist science is one of the few 'post' prefixes that sounds positive. Certain scientists are expanding their philosophy and method. This should qualify the idea that quantification is the primary or sole way to truth. Qualification of quantification in favour of qualities or qualia should happen. A theory of consciousness may marry knowledge of qualia with quantum theory to reduce the classic, quantification trap of an Obsessive Quantitative Disorder! We need balance between useful measurement and recognition of qualities. We should recover a sense of personhood and humanness to defend the whole individual. Obsession drives our world. The best and worst of us are driven by obsessions. We must not be entranced by measurement to the detriment of phenomena. We must not mistake the immeasurable for the worthless. We must not treat qualities as immaterial or unimportant because they are physically immaterial or insufficiently identifiable by instruments. We must not reduce others or ourselves to objects. If we become aware how attention is marshalled towards processes that are not as significant as the qualities of personhood, we might become more reluctant to over-value the mechanistic.

Good evidence should be good evidence. Numbers, information and data are critical for good decision-making. Perhaps it is ironic that William Stern was involved in the evolution of intelligence quotient tests. However, appropriate quantitative information and accurate statistics with sound methodologies without ulterior purposes are important when one wants to extrapolate and interpret accurately or construct models. The quality of evidence may be compromised by the quality of its communication, representation or even misrepresentation. When metrics become tyrannous and are elevated to idols with god-like status through a cult of information and priesthood of experts because of the scientific religion or ideology of scientism, perhaps informed by a moral relativism adapted to the purposes and propaganda of data producers in a context of confined freedom of expression, then storm clouds build. In that vexing vortex I suggest some re-alignment, looking beyond traps of the materialist blinkers. We should be willing to widen our way of interpreting the world, incorporating lessons of traditional and local knowledge. Obsessive quantification is familiar and satisfying in the face of uncertainty, just like magic. As the quantum world indicates that ostensible certainty of the classic scientific approach is less clear than appearance suggests. When the natural, physicalist, materialist models or methods are properly kept within appropriate channels they may work very well. The quantum world however now seems conceptually more similar to magic. We must be able to accept complexity, uncertainty, unpredictability and mystery. If we seek to make models of control solely to reduce uncertainty, we may engage in magic and not science. The unified comprehension of quanta and qualia as aspects requiring an observer and the requirement of personal consciousness reinforce a participatory universe not limited to slides, models, specimen, formulae and maps as absolute truth. The reason why we should focus on a proper balance between qualities and quantities, classic natural science and quantum science, mysticism and materialism is so we can understand comprehensively the complex nature of our consciousness embodied and incarnated in us, available to us and requiring vigilance lest we be compromised in our ability to exercise its full potential. Roszak indicated that good magic clarifies and bad mystifies. The mathemagicians may make models that allow digital horcruxes be made but such sigils will never contain the human spirit. Material seeking of immortality and deadly desire to defeat death may however indeed constrain spirit and the nature of the human race.

Servitude, Deconstruction, Incarnation

"And the light shineth in the darkness; and the darkness comprehended it not."

John 1:5 *Bible: King James Version*

Perhaps there is a loose link between the critique in *Discourse on Voluntary Servitude* by Étienne de La Boitié (1530-1563) of the mid-16th century and the remote efflorescence of a critique of the technological society a few centuries later in the same area. The argument in the idea of voluntary servitude was based on people. The later argument was more based on technique and technology. Dominance secured through superior technology does lead to elite control. Dominance by technology itself, with associated technique or governance machinery, represents qualitatively different phenomena in some ways. Such arguments are united by comprehension of the role of the individual person in allowing such tyranny emerge and also resisting it.

The two major French critics of technological society, Charbonneau and Ellul, are perhaps linked by the idea of incarnation according to some commentators. There is no doubt they were influenced by a Christian background. The suggestion is of the importance of accepting that we are who we are. There could be two sound bases for this in my view. Firstly, we have evolved over millennia to our present state. That the people who proclaim evolution always want to move on from it or replace natural selection with artificial selection is astounding in one way. We are adapted to this planet. Our problem results from moving away from our relationship to it. The commitment to more technique and technology must mean that the technosphere grows at the expense of the biosphere and ethnosphere. Secondly, the spiritual sense of some traditions such as Christianity emphasise the divine descent into flesh as a critical part of theology.

Charbonneau and Ellul operating in the South West of France point to a tradition which echoes the American Transcendentalists and perhaps mountainous metaphysics of John Muir and the Sierra Club which indicated hostility to technological dependency and prefigured the ecological movement. Charbonneau also criticised Teilhard de Chardin and so went against the grain. Charbonneau underlined the significance of the person and spirit in an embodied context in nature. Both rejected left and right approaches. Their spiritual truth seems consistent with perennialism. Their critiques anticipated later ones of quantification and progress and were in favour of more spiritual frugality. This was advanced in the 1930's onwards. Charbonneau argued in favour of the remnants of 'peasant' culture in France and embodiment in the environment in a real relationship based on taste for him.

Charbonneau and Ellul were part of a more widespread movement of personalism. There were Christian and peasant influences conjoined, Catholic and Calvinist with Marxism. A critical idea for Ellul was of the real presence around him. He experienced the Divine as a mystical experience and was converted in that way. Both of them saw the significance of embodiment. Embodiment and incarnation is presence and reality. Ellul identifies touch and presence in the life of Jesus. He explains that the life of Christ involves being in flesh, being incarnated. This is an important part of the spiritual quest which must be accommodated in mysticism. Those who deny the flesh may also deny the significance of nature and focus only on parts and not the whole. In the whole is the Divine, immanent and transcendent. Charbonneau thought that churches and the Catholic Church were cowardly in the face of new orders of governance. The Aquitanian approach had a tradition of freedom. Charbonneau saw Teilhard de Chardin as totalitarian-tending.

There is a very dangerous and insidious fashion with some thinkers at the moment in the general academy of philosophy. It purports to be an enlightened search for the 'truth' that involves a deconstruction of the person or personhood. With a straight face a 'philosopher' may say that they do not exist, they are not there, they are not the person who we believe them to be, their identity is non-existent and so on. On the face of it, this view appears consistent with a whole range of philosophical streams and the case can be solidly made on paper or in argument. When made by materialists or even post-materialists, such a view may be calculated or have the consequence of deconstructing the person. This approach would not be a problem, if there was not a strong, ideological force which actually seeks dissolution of humanity and personhood. Certainly the notion of transcendence involves a rising above the ego and personal identity. However the effort at transcendence is an affirmation of spiritual consciousness and not a denial of it. In the context of a deconstructing, dispiriting discourse, such contributions that may appear enlightened are really serving an actual ideology that seeks dissolution in practice. Beware of those who sit before you and tell you that they do not exist, that their name and identity is meaningless and they are something else. They may contribute unwittingly to the promotion of a policy which denies your existence and fails to see you as a person with dignity or reality worth considering. You will be a bit to be managed, moved and manipulated and your existence becomes meaningless. Remember for some people that it is about constructs, deconstruction and reconstruction. That is why there is so much talk about 'sense-making' which is a concept entirely consistent with construction or re-construction or building back better. Part of the re-building and re-constituting involves re-building you. Be careful with the apparent playfulness of such approaches. While it may be Russian Roulette for some, for others all the chambers are loaded. Pragmatists should ask, what are the consequences of such a viewpoint? Deconstruction of the individual person is consistent with a hostile tech-takeover of your life.

Incarnation was also supplemented by the idea of presence. We are here and now as far as we can tell. Others want to cast a spell to suggest we are not and then perhaps create a self-fulfilling prophecy by ensuring that outcome. The person must be present in a real world with all their inherent and tacit knowledge that is trustworthy. Deconstruction can become a technique of the scientific or philosophical slaver to serve a cut corpse on a salver. Self-hatred crept into academic disciplines and became a programme or policy in certain commercio-scientific contexts. The assault on personhood is another point of attack from the Enlightenment and Empire on nature. We are told that the megamachines that drive the madness or drives with madness are going to save us all. We should trust people whose thinking allowed mass systems evolve.

Those who helped dig up the earth and pollute water and atmosphere are discovering ethics. We are not to worry that they will bet again on the losing square. We are to thank them for increasing population and then ask them to solve the problem. We are to sit back while they drive us to destruction. We are to celebrate the wisdom of their method supposedly altruistic and restrained. We are to ignore millions dead as a result of their machines. We are to feel safe with anxiety they create. We are to blame the Church of 500 years ago. We must throw out all spirituality and religion which are merely superstitions. We show ridicule to the supernatural despite science's increasing discovery of more senses than empiricists hitherto allowed. We are to trust people who have often dispensed with ethics. We must ignore writings by scientists who declare their desire to imprison us as docile lesser creatures. We are to forget scientific racism and allow behaviourists foment discord with newly-minted versions. We must take a new gospel announced by neural correlations as the final, absolute comprehension of consciousness from disciplines that have no idea of what it is and even whether it exists. We are to fall for the shiny lure again from some sweet promise that masks an ill-conceived jamboree for scientific corporations. We must cede everything before they compel the wonder of flesh and blood by a black mass of inverted logic into a machine. We must accept disintegration as integration. We must be bound as if by evil spirits by the conjurer who chants in the dark protected by a circle of power and cloaked to conceal and glorify. We must allow the dominion of the Empire of Scientism render us into slaves pending our disintegration as whole beings or organisms.

Thus the philosophical exploration of personalism (which may also be deconstructed, hijacked and re-constituted) or our inherent knowledge, instinct and intuition about whom we are as people with subjective agency and intentionality are very relevant. The spiritual and mystical studies that re-affirm significance of the person are important. Otherwise we will relinquish our very spirit or psyche to a falsehood. In this narrative of dissimulation we face dissimilation from humanity as we are assimilated into an apparatus of control. You will be made different daily so you can and will be shaped into subservience to be assumed into a network machine. Some apparent champions of liberation want us to get out of our head, whether through mind-altering substances or technology. Getting out of our mind to our spirit will help us but lotus-eating will not. Neither will our collaboration. Arrows on the market floor, electronic access, tracing methodologies, mass surveillance, robots and automation indicate that the technosphere is growing at our expense. The world is being re-made for machines and we will be managed, manipulated and manacled in a machine world. This occurs because we know not who we are and have lost confidence in our personhood and abandoned our commitment to others. In failing to recognise our own consciousness we cannot see others. Thus atomised and alienated by attending to algorithms of the tech-Leviathan and hounded by a Hydra of technique, we will become pets and begin to become shadows of the great shadow of psychic control. We are flesh and blood and within that context possess functions which distinguish us and make us more angels than beasts. We are spirits as well as physical creatures, we are embodied, incarnated.

There are some thinkers who discovered our true nature and tried to turn those personal perceptions into philosophy congruent with modern physics and ancient mysticism. As mentioned above, one neglected thinker is Benjamin Blood who influenced William James. Benjamin Blood explained the significance of anaesthetics for coming in contact with true consciousness from which all true philosophy flows. He probably anticipated the significance of the later psychedelic movement, operating-theatre NDEs and general mystical experiences occurring likely through suppression of active functions of the brain's filtering mechanism. In addition, Blood explored ideas of a 'pluriverse' which may anticipate ideas of hyperspace proposed by scientists and psychical researchers such as Bernard Carr.

As a sunflower seed contains the being of its own future within, it still seeks the sun as the external force that interacts therewith to produce through synthesis its full potential in a cycle of continuity, constrained by the external container of shapes and principles of growth and form. Now we want to relinquish that superintelligence of natural selection within us for the supposed one of ratiocination that is unable to perceive its own limitations and has a history of destruction and ugliness. With a re-invigorated idea of consciousness, personhood and their qualities we have shields to protect us and a sword of light to shape a new universe. Love of machines that replaces love of people is an attitude that has gone too far. Excessive extrapolation from theories of evolution has led to misleading ideas of competition that threaten now to directly cannibalise us. Tech-billionaires benefited most from the crisis and the monster they represent spreads.

Ante-Humanist Metamorphosis to Animasphere

"After that he appeared in another form unto two of them, as they walked, and went into the country."
　　　　Mark 16:12, *Bible*: *King James Version*

Long, long, long ago people realised they were people or humans. If we were raised with wolves it might be difficult to identify who we are. But generally humans know who they are and that they are different. The idea that the concept of humans is a mere invention or construction is part of the type of logic that affects the body politic like a cancer or virus. While there are specific trends in history wherein the nature of humanity was spelt out because of an attempt to comprehend humankind in a theological sense and to work out schemes of individual protection against arbitrary power, such writings never invented the human nor the concept thereof, nor did science. The ante-human, the pre-existing human perhaps suggests an ante-humanist position or recognition of the perpetual nature of humanity independent of naming or classification. To establish the human as a mere construct is to set up a straw man to turn us into straw dogs. It is strange that the hair-trigger sensitivity promoted by the propaganda system and corporations is so tolerant of a bulldozer mode of thinking that threatens to eliminate humanity itself. This suggests that such Punch-and-Judy shows are part of the circus of distraction and the devil of division. You are not a fabrication.

The technosphere is about monetising, systematising, homogenising and standardising the world. Driven by curiosity, technique, technology, commerce and science in a technocratic society the technosphere grows at the direct expense of the biosphere and ethnosphere. If something like the 'animasphere' is enlisted then the opportunity for balancing is possible. There would be benefits to us. By the animasphere I refer to a concept I have advanced to indicate the imaginal, mental, but more importantly, spiritual space representing consciousness itself manifest within us. It is the sphere of myth, legend, poetry, higher music, magic and the collective unconscious. It is the sphere that mystics go to and through. It is the sphere of psychenauts. The anima and animus are part of us all according to Jung. Anima indicates the breath or spirit. It is consistent with the transpersonal sphere of imagination that some of the personalists indicated and mystics like Swedenborg, Blake and Yeats celebrated. The influence of mystical Kabbalah also requires the imagination. Animasphere could also accommodate respect for the animal world. The word animal is etymologically based on the Latin word 'anima.' Denying the animality of the human and failure to recognise our physical nature and connection with other beings has been a significant issue that promoted posthumanism. It is also interesting that the negative word 'animosity,' used to be a positive word suggesting courage in the past. We need to get serious about concepts. Some posthumanists for example argue that humans before humanism (which I called ante-humans if anything) were posthumanists. It requires that sensible people respond to public debates and get involved in rational and reasonable discussion.

The animasphere might be an antidote to the narrowing humanism that contracted after its original flowering in the Renaissance. By focusing exclusively on the tool of reason, humanism has sacrificed the very elements that made it. Concentration on reason has contributed to the excesses of capitalism and scientism. If the rational, reasonable, materialist focus is creating a technosphere characterised by networked transhumanism then an idea of the animasphere may counter-balance it. A full embrace of the animasphere may also encourage solutions that emerge from higher consciousness. The influence of poetry in the origin of humanism should not be ignored.

It is fundamentally necessary to give correct diagnosis to a problem before one seeks to solve it. The first point then is to make the case that the direction of technology and use thereof increasingly to control us, is an existential threat with a dark and difficult final stretch before us. Once the grim analysis is made then the bright approach may be advanced. The light is associated with a recovery of elements we once had and then discarded. Reclamation of our imaginal and spiritual sense on an individual level is the only antidote to the mass-mind of machines. We have become entranced by an image of ourselves fashioned from fragments of our base mental elements and emotions. Material individualism is subjected to easy manipulation on a mass scale through a black magical strategy that led us into a dungeon. The need to recover and rehabilitate is not just a reaction but realisation of who we are. Our problems emerge from having lost our way and succumbing to machine, instrumental, technical, technique of materialist control through reduction, quantification, disenchantment and dispiriting.

The difficulties point to the absence that caused our problems and predict and promise a better path by advancing on it again. Transcendence that we should seek is spiritual and enlightenment is of mind and spirit. The mission of the Enlightenment has been to substitute spirit with a simulacrum and the result of the success of that endeavour will be an inevitable extinction and an ineluctable decline. Find a robust critique that works or you will fall for anything you do not have the time to examine properly. There is an idea of 'meta-humanism' which is yet another movement. It is clear from the pronouncements from its proponents that this is not a mere description but a political or ideological movement more than philosophy. But old-fashioned metaphysics is useful as being prior to epistemology. Transhumanism must involve physical or material change with consequences in the natural world. Transcendence in spiritual terms represents a journey in the animasphere which has indirect consequences in the physical world. The metamorphosis we need may be provided by embrace of the animasphere concept and reality or some of its established equivalents. Metamorphosis refers here to change at a spiritual level. Perhaps form might at some time be affected just as a smiling face may develop nice, signature smile-lines. But one starts in the spiritual domain while transhumanism proceeds exclusively in physical, hi-tech dimensions. Metamorphosis also occurs at a community and societal level if successful. You are not a mere observer in society. You are an active agent. You are a person. You have views and interests. Others who have views which are often strange will project them on you and affect your life if you snooze or settle for the packaged schmooze.

The person or individual who then appreciates their psychic or spiritual potential can facilitate others not least through recognition. Perhaps we need a move from anti-humanism to a prior ante-humanism in some ways. Any such return might have to see the world differently, perhaps in a re-invigorated animistic way informed by the better observations for post-materialist science, retaining contributions that did ennoble society. Concentration of humanism has formed to a cone point, away from a wider holistic base. Ante-humanism is a term I would use to indicate a condition beforehand in the Western mind or indigenous peoples which involved a human integrated into community. Humanism has been purified to a point of pointlessness in some ways by science and reason. The later dispiriting movements removed the foundation on which constructions were actually built and grew. Humanism emerged on assumptions that were embedded in the community with a sense of the cosmos. It is a conceit of the rational, reasonable, materialist mind-frame that everything sprung into being from a tool or merely a disposition revealed once religion had been attacked. There is evidence of nostalgia for the Middle Ages in personalists like Charbonneau and Berdyaev. There are other tech-critics that suggest we should move back to such a stage technologically. There are dangers with utopianism and arcadianism merely wishing to regress. Such turn-the-clock back responses can be seen on the left with the Rousseau-loving Pol Pot (1925-1998). Talk of the 'Great Reset' seems to suggest a new era of imposed tech-supervised frugality and serfdom. New serfs or surfs will not enjoy any rural idyll. It is one thing to think of looking at a time in the past and another to begin re-enacting it.

The reason why spiritual consciousness is being ignored partly is because it has not been employed and demonstrated significantly. This is the context in which prophets arise. If they do not arise their absence will seem to prove their non-existence. The story of Elijah comes to mind. However, it is not more violence against people but rather an engagement with the world we need. But in order to provide a critique, one must present some counter or positive image. Transhumanism and the techbondage of the TechBondAge that will rise with the Empire of Scientism, is related to a process of dispiriting. It is only thus by re-animation of the world and our minds that the perception of possibilities and potential will then arise. Dispiriting is only solved with re-animation. Spiritual consciousness provides a defence to the slings and arrows of life and a source of solutions to problems encountered. For some, the idea of 'animism' may allow reconciliation with indigenous perspectives. If we manage to counter-balance the tech-hypno-trance and paralysing spell that writes our species suicide-note in code, then we still must encourage spiritual consciousness. We face the continent of contingency in the Empire of Scientism which is colonising us. It can establish itself if we do not recognise our personhood, commit to spiritual evolution and seek to self-actualise. Our freedom is within us first. The outside world is a manifestation of our internal state. We allow a drift into servitude. We permit bondage of technology. We must be careful to ensure that the thin edge of a wedge of technological restoration of the human body is not then followed by a technique of network transhumanism. Our failure to liberate our own spiritual consciousness will lead to a takeover and assimilation of us into the Machine.

You are powerful potential. However, keys to unlock that potential require consistent discipline and rational self-examination. Will-power that accomplishes things in the world comes from the ground of being. It is only from understanding of our own nature that we can contribute to society. Through recognition of ourselves first, we can recognise others and develop empathy and compassion. Empathy and compassion are powerful forces to unleash. Love has been devalued through commodification and the black magic of pop music. Beyond the illusion of that inflated term is a real force. That quality is immeasurable, beyond measurement, beyond reduction. It is a force of consciousness. It is the highest consciousness. Blake said that science is the 'tree of death.' Transhumanism, as part of the scientism ideology will be a magical transformation of us into matter, things or a different form. This is what many legends and myths warn about. We are meant to be in the form we are and accept and maximise it consistent with the biosphere, subject to restoration to normal or legitimate expectations where needed. The personalist philosophy from Scheler through Germany, Wojtyla and other Polish and US philosophers, MLK and through the French tech-critics provides a possible direction. The theorists that have been focused on the person as a whole organism have often celebrated the significance of creativity and the imaginal as central defining qualities in the individual human person. Kurt Goldstein also saw the process of healing as an act of emphatic imagination. Smiling infants recognise their fellows and share their mutual correspondence. Spiritual consciousness will combat totalitarian tech-consciousness. We must value our person as humans and invest in our spiritual evolution.

This is my take on personalism. The person is the basis of knowledge and the starting point is about them. Within us is a need to develop vertically toward the higher consciousness and good, transcending ego and through participation in the world recognising the personhood of others. This requires that we respect nature and our relationship to it as people and accept that we operate in natural contexts as far as practicable. Perennial philosophy indicates a path and an experiential route not dependent on church machines. Accepting that we have evolved physically and spiritually, the choice about 'evolution' cannot be made by any small group of activists artificially. Technological change of the body is not evolution save by sacrifice of other, existing potential or qualities. Imposed mass-evolution through forced network transhumanism would be enslavement and genocide. Destruction of the environment for benefit of the technosphere is 'ecocide' for some. We must see the world as it is and assess truth.

Truth is not merely a propaganda statement nor what power says it is. There is something good or bad that can be reasonably identified with interiority and conscience. From a pragmatic perspective, philosophies of personhood have consequences. Bad philosophy kills. We must be subjects who comprehend, understand and empathise. As persons, we interact with the world but there remains the part in us that is the seer. For some this is universal consciousness or God. Therein is our need to experience ourselves and to make decisions seeking to promote the good or growth. Transhumanism seeks to deny that which is there and pretend that a maximisation of life tending to physical immortality and morphological freedom is possible. The price for falling for this lure will be huge.

The persona may mask but the seer remains behind. Network transhumanism will seek to obscure the seer and the person in the deepest sense. This is the strategy of collectivism to control the individual. Personalism and the process of spiritual evolution is the other path away from domination by the Machine through resistance and solidarity. To escape nihilism and alienation of machines through technocratic, instrumental, utilitarian philosophy, we need a personalist philosophy. Most of all remember - we know we are. You are not you because you think but because you are conscious. You are. You are a conscious person. The dichotomy is not to be or not to be but to realise that you are or not. While you are part of the world and biosphere you are a sovereign being. Aware of this we can seek to act like conscious and sovereign beings. If we do not we will be assimilated by an authoritarian system. Eisenhower's Farewell speech in January 1961 included this famous line.

> *"Yet in holding scientific discovery in respect, as we should, we must also be alert to the equal and opposite danger that public policy could itself become the captive of a scientific-technological elite."*

It was not curiosity that drove scientists but government contracts. Through such contracts, scientific corporations predicted to run the world a century ago came into being. This capture of the pump of public funds was not the end but the start for them and maybe the start of the end for us. Any totalising tech-solution will increase our subjection and objectification until we let ourselves become things.

These legal persons with enormous wealth can then vampirise democracy, zombify scientists and combine transnationally to create the Faustian dystopia feared by some. It is only through knowledge of ourselves and exercise of vigilance that we can respond creatively to a clear and present danger. A solution to de-personalisation is to re-affirm the person and to deal with de-humanisation is to re-iterate the human. Within our personal, collective unconscious and animasphere are archetypes that can assist. Simpler than that, we can simply value and protect personhood, vigilant against erosion by convoluted attacks with the aura and arrogance of authority. We can think for ourselves in a fashion that is not merely a rehearsal of old philosophy. We can seek to recognise our personhood and others and be personable to find common humanity. We can be more aware of the perennial trajectory of spiritual evolution of consciousness superior to reduction and translation of humans into machines. De-biologicalisation is diabolical. However, spiritual evolution must occur and mystical tradition must assert itself. It cannot be some soft-focus plea or platitude for peace in chaos while we disintegrate. Our spiritual consciousness must improve and grow and transcend compartments that culture forces us into. We must escape from narrow doctrines that allow us be divided. When we are made to be preoccupied with spurious policies and invented wrongs and hair-trigger opprobrium in the technium by managers committed to its apotheosis - our decline will be rapid. The solution is not to give more power to artificial persons like companies, corporations or even cyborgs, androids or mutants. The solution is to unleash your divine personality. Watch for demands that you give your personhood to something else.

Erwin Schrödinger (1887-1961) lived in Dublin for 17 years. I went to Trinity College Dublin where Schrödinger delivered his famous lectures 'What is Life?' in 1943. He was a thinker who understood qualia and distinguished scientific description of phenomena and the phenomena themselves, in this case the quality of ultimate perception. He understood how biology would develop. Otherwise, we may be seduced by a reduced picture of ourselves no matter how sophisticated. A sophisticated argument is an idea that can have negative implications of sophistry. We have ways and methods that have been deeply explored by philosophers like Husserl, such as phenomenology. His 'lifeworld' allows us apprehend consciousness of the internal and intentional as it is, rather that engage in explorations to be reduced more.

In all this, the idea of evolution is critical. However, evolution refers to gradual development and adaptation. Folly of planning and mentality of excessive centralised organisation should warn us against any idea of general plans to transform our species. Artificial intervention is not natural selection. Arguments that prosthesis is like a tool for humans are not persuasive. Bearing in mind the disastrous role of central planning in the twentieth century, whether on the left or right, we should not embrace it for our body, mind and spirit. Such spurious 'evolution' is based on a caricature of the human person. Science cuts the person to fit the picture and so precludes the process of natural evolution and most importantly spiritual evolution. There has been a certain failure of spiritual evolution. Such a failure in evolution of spiritual consciousness has caused our predicament. The cure comes from where the poison is. Myth tell us not to fall in love with a surface reflection that has no existence and not to trust in technology or knowledge above all else. The word 'transhumanism' has many anagrams, some maybe consistent with its objective like 'antihuman,' 'inhuman,' 'harms' 'huntsman,' 'artisan', 'marsh', 'satan,' 'sham' or 'smash' but also 'manna,' 'mantra,' 'shaman,' and 'ashram.' We can all seek to see what we want to see and transform that which we want to.

We should recognise ourselves as whole human persons with vital force within us which is never mere mechanism. We are composed of the most fundamental consciousness, incomprehensible to science. We are whole beings greater than the sum of our parts. Yeats had a conception of magic that he espoused in his essay on 'Magic' in 1901. He was taken with an idea of the 'Great Memory' or mind which we could all access through symbols. The Great Memory is perhaps consistent with the animasphere. The magic of Yeats was quite consistent with mysticism. Many psychologists have argued that we draw power from creative imagination. We must do so now. If you oppose the change of human to machine, you are immediately labelled as anything from a mild 'bio-conservative' to 'deathist' to daft and unfortunate rhetoric about being 'suicide bombers' of people's expectations. You will be forced to re-evaluate who and what you are. You may become less than an android. Better be H.I.P: Human, Individual, Personal. Protect personhood and resist dispiriting. Do not fall into group-think that suits propaganda of the producers of our society. It is only when we experience a personal metamorphosis that we can contribute towards a more general one.

Human demise can be seen from the time of Darwin. It is worth recalling words of Samuel Butler (1835-1902) in 1863 in an essay 'Darwin Among the Machines.'

> *"The views of machinery which we are thus feebly indicating will suggest the solution of one of the greatest and most mysterious questions of the day. We refer to the question: What sort of creature man's next successor in the supremacy of the earth is likely to be. We have often heard this debated; but it appears to us that we are ourselves creating our own successors; we are daily adding to the beauty and delicacy of their physical organisation; we are daily giving them greater power and supplying by all sorts of ingenious contrivances that self-regulating, self-acting power which will be to them what intellect has been to the human race. In the course of ages we shall find ourselves the inferior race. Inferior in power, inferior in that moral quality of self-control, we shall look up to them as the acme of all that the best and wisest man can ever dare to aim at. No evil passions, no jealousy, no avarice, no impure desires will disturb the serene might of those glorious creatures. Sin, shame, and sorrow will have no place among them. Their minds will be in a state of perpetual calm, the contentment of a spirit that knows no wants, is disturbed by no regrets. Ambition will never torture them.*

> *Ingratitude will never cause them the uneasiness of a moment. The guilty conscience, the hope deferred, the pains of exile, the insolence of office, and the spurns that patient merit of the unworthy takes these will be entirely unknown to them. If they want "feeding" (by the use of which very word we betray our recognition of them as living organism) they will be attended by patient slaves whose business and interest it will be to see that they shall want for nothing. If they are out of order they will be promptly attended to by physicians who are thoroughly acquainted with their constitutions; if they die, for even these glorious animals will not be exempt from that necessary and universal consummation, they will immediately enter into a new phase of existence, for what machine dies entirely in every part at one and the same instant?"*

We are the people who must assert our inalienable right, destiny or duty to maintain personhood as a manifestation of Divine consciousness. As moguls meet in Sun Valley and elsewhere to rule us, do not expect democratic interests to be promoted. Bernal predicted the human zoo. Get ready with more Zoo TV, CyberZoo, Zoom and Zoox to reach that space. Many writers realised the implications of technological advances just as Mary Shelley or C.S. Lewis did. However it is a mistake to believe that they saw this as an inevitable phenomenon of transmutation. Rather they were indicating our need for choice.

But it is not a matter of what some thinker has said, nor merely an issue of philosophical compartments and how something matches with the past. Each individual has consciousness which allows personhood, selfhood, spirit and our inherent humanity which is self-evident. It is only when mind-tricks are played that we can be talked out of this heritage as if in a hypnotic trance. We should recognise that as individuals and as one race of beings with common humanity we are under assault. We are under attack from a deep, materialist philosophy and an associated ideology of scientism, a dangerous combination of technique and technology and particular thrusts of practices and policies such as transhumanism. These are not academic or hypothetical issues but identifiable forces in a contest for control of people and the spirit of the human race itself. Effective marshalling and magnetising of opposition to the power and potential of propaganda and perpetuation of the Machine or technosphere, requires a robust critique and sound analysis of the contemporary cultural and commercial landscape. As robots roll out, room for freedom diminishes and the air of liberty is thinned, we will soon discover that wishful thinking and avoidance is no longer an option. It may be a zoo or, as Harari indicates, we may be like domesticated animals such as cows or dogs to Butler. Or we may be turned into something that functions as a node in an almighty network. Every time we have been sold a wonderful network, we have embraced it, oblivious to its effects. This is our final embrace. Many spiritual and esoteric traditions are about one struggle. This is a perennial one between sacrifice of contingent spirituality or commitment to actual and promissory material or matter.

It is a contest between miracles of technology and those of the spirit. One will force us to be automatons in an automatic world of machines and the other will foster autonomy. Another link between magic and magicians and transhumanists is that they have always been interested in mechanical automatons. Now we are to become such machines under spells that suggest we will be improved. Magicians will seek to use the hi-tech network as a cloak to cover consciousness and create a false substitute for spirituality in a simulacrum that purports to serve needs of the spirit of the human race. Standardisation of spirituality will be an effort to corral further the flock in a mass delusion of reality. When mock, pseudo-spirituality is fashioned it can further oust a traditional, genuine and real type. That threat requires a genuine mystical awakening to survive personally and protect others. Well-meaning innovators like Tim Berners-Lee make clear that social principles should reflect the systems he has driven forward. Many see cryptocurrency as an enlightened economic or financial alternative. All such approaches will necessarily expand the technosphere. It is time to take a dose of scepticism and critical thinking and draw on a wider range of intellectual sources. Freedom promised by technology is illusory, absent genuine theoretical or philosophical anchors. In *1984*, the Party constructs reality and disabuses the subjective, human, individual person that any objective reality exists. We are in the stage where the concept of the person is being rubbed from the blackboard of memory. Sum ('I am' in Latin) becomes a data sum imprisoned through blocks and chains, linked in to final dissolution. Transhumanism akin to magic will prove to be destructive to humanity.

We need an antidote to relentless pro-tech analysis as part of response to the dynamic world. With technological expansion, loss of social intimacy and more alienation we need to re-assert our personal sovereignty. Imagination, the imaginal world and animasphere are critical. Ellul rejected Marx because of the latter's failure to respect the power of imagination. Imagination must triumph over ideologies in a pragmatic way. We must rise to the new challenge of technology, not rejecting science but making sure it serves humanity and not that we serve it. As you notice more robots in your environment, encounter more restrictions on freedoms and see how technology controls behaviour, think carefully about the value of turning from a person to a thing in the internet of things. The philosophy of personalism represents an attempt to respect the human, individual person as more than mere physiology or a collection of diverse, psychological attributes. We are persons and subjects and not objects or things. We face mysteries that are incapable of simple resolution. Like with quantum physics, we are participants who influence and are influenced by forces they deal in. We must engage to combat with gifts of hope and gratitude against despair and dispiriting. We must make sure that we are not really reduced by reason and objectification and instead see the greater force of love beyond the material. Some personalists talk of a process of active love which is not mere fantasy or purely self-interested. Reclamation of the power of love as an active phenomenon away from disillusion of the technosphere propaganda system is critical. The imperative of love is the only force that will override the imperial power of scientism and associated rackets based on its awful power.

Nearly a century ago Bernal adumbrated the dreadful future based on scientific control with a breakaway civilisation, space colonisation and scientific control by stealth and transhumanism. He was nicknamed 'Sage' or The Sage. What a coincidence that a key, power group in the UK is the Scientific Advisory Group for Emergencies (SAGE)! The docile public would become potential for experimentation in Bernal's opinion. Others said we would live in cities under glass domes, or otherwise be concentrated. Now the glass dome experiments will begin as the robots proliferate, nano-bots expand and a scientific ark is established to escape the mess they intend to make. Maybe they can have a laugh with some Noah acronym. We need a spiritual sense both to deal with the madness or destructive magic and to create a better future than the grim dystopia some scientists have prepared for us. As the West retreats from a war it could never win, exposing the bloody corporate racket that had been camouflaged in obfuscatory strategies and spurious justifications, it is worth remembering how dependency can be created by systems and support and then withdrawn by those who use the cover of propaganda and purport to exercise the power of authority for the supposed common good. The deliberate breakdown and withdrawal possibility in order to reconstruct is one suggested by Ayn Rand. The gross, physical being of our body will be reduced by deduction of all that is unique about us to a reduced net of us. Not only are we trapped by the net, as its very name signals, but we become integrated into a net so we become a net result incapable of remembering our past and potential. Prosthesis really means to put something in place. It is us who will be put in a place calculated by the Empire for us.

We need an appreciation of something like the animasphere to help shape our imagination and release the valves of emotion and allow our soul flight. Appreciation of the animasphere may allow conscious escape from the technological cage in the shadow cave of the captured and bewildered of us. It suggests the *anima mundi* and related ideas of active, vital and alive forces in us, nature and the universe. It allows a richer conception of the planet as a living force while linking into the mystical traditions. We must seek genuine art informed by a spirit of elevation and reject the false industrial art of much rock and roll and tv where pop stars do propaganda. These have been mere lubricants for the Empire of Scientism and we have been fooled, soothed, infantilised, lulled to sleep, hypnotised and put under a spell. The drone of the dull machine music has helped turn us into drones to be managed by drones and turned drowsy for the androids. Edgar Allan Poe published an essay called 'The Man That Was Used Up' in 1839, about a hero who turns out to be predominantly made of prostheses. These issues are not new save for the immediate threat of transmutation of humankind into cyborgs, diluted and reduced in technical proliferation of androids, robots and pervasive networks as part of a perverse dream of progression whereby we are gradually linked in, fixed with blocks and chains, captured in the net, put on line, stuck in webbed dreamland having bitten the apple from the tree of science and found ourselves confined as domestic animals. As humans we are not mere constructions, nor made up. We must have courage to explore our full potential. There is no problem with space exploration or technological enhancement but with scientism, corporate-impose, network transhumanism.

H+ is the symbol for transhumanism. H+ is Spirit-. Merchandising of human mechanisation is likely to be a mass net subtraction with exceptionally contained benefits and huge costs. If humans become devalued physically we become trash burdens to remove. Look at what some of the proponents are saying. The argument that it is about individual choice is inconsistent with reality. It is interesting that the name Eichmann has the word 'machine' in it. The human has never fully appreciated the impacts of machines on it. Berdyaev was right to indicate how humanism went wrong during the Renaissance. When humanity focused on the machine and power to create instruments to give more power to glorify the element of the disposition that dwells thereon, it was partly rejecting the spirit and the inherent idea of existence of higher consciousness. If we then adopt transhumanism or posthumanism as an attempt to bet again on that mistake we will get more of the thing proponents claim is the problem. Further distillation of materialism concentrates the madness further. The only curative hope is that spiritual evolution, spiritual powers, magic of a mystical type consistent with perennial philosophy and some cosmopolitan personalism that is pragmatic can provide courage as antidote to this poison. Mirandola can be cited as being consistent with the superhumanist trend. This may be correct or it may be that materialists miss some mystical arguments. We should look at some of these but learn things that work today, help explain the corrective and ultimately evolve spiritually. We are not defined merely by our capacity to calculate and if we so believe we will recreate a world based on this attribute alone which must then reduce the human to fit the cut.

I have not engaged in some of the more outlandish claims such that transhumanism will make people equal. It could indeed seek to do so in a creepy cookie-cutter way. We must be careful to know ourselves. We should be aware of the mental robot in us that Colin Wilson described. The robot is a projection of a force within us. It too can be useful to perform mundane and necessary tasks, but it can also take over and displace openness, joy and the unpredictable nature of vital forces within us and the world. We must go back to the source in us to see the forgotten aqueduct of consciousness through the fog. Otherwise you face a world where the driving forces in science claim that the person, self, free will, spirit, mind and consciousness do not exist. As they do not exist and are merely illusory, the protections offered thereto are also fictions. We can expect the same treatment as the poor bats being experimented on that are claimed to be the scientific origin of our present scientific dictatorship. A tract does not have to be attractive in its argument to be true. We are being riveted as we are colonised individually to be clownish clones that become drones drowned in data and attenuated by an anaconda of control before our ultimate assimilation through our final acquiescence. Let me put the problem simply. *The mind that seeks predictability and obsessive control of others because it is more interested in things than persons has achieved the concentration of power in the technosphere necessary to project that totalitarian paradigm onto us and to proceed with this politics and policy to the highest degree by using implantation of devices to turn us into machines or things to be managed, firstly as slaves and pets and then as constituent parts of the very network itself.*

You may think that the new world will be shiny, clean, convenient, fast, full of machines, nice, easy-going, push-button, cosmopolitan and free. You may have a vision of your children floating among the stars, caring for each in a new way. You may seriously entertain (and entertainment is the right descriptive concept) a future magical place of scientifically, measured and enlightened beings. However the drivers of this vehicle told us what they wanted. This was made very clear by the Italian Futurists in the first decade of the 20th century. Their vision either influenced or accorded with the fellow imperial tendencies in London at the same time. Italy is a natural home of new imperial tendencies. Legal cases associated with Propaganda Due or the Mafia should caution against believing what you see on tv and demonstrated how covert much governance has been in that state not to mention the separate Vatican. But the striking thing about much Futurism was that it even left Nietzsche behind. It celebrates death, power, speed and the machine while attacking kindness and concern. It thus seems to invert true ideas of humanism that have been reduced and subverted by its critics. The obsession with immortality is clearly a reward for the Faustian assault on the future. It is not motivated by a love of humanity but by hatred thereof and that machine misanthropy transmutes today with a soft-focus sales pitch. Values that will thrive in this metallic paradise are associated with amorality and a Romantic association with sensibility. People have failed to see the link between the Romantics and machines. The attraction to the awe-inspiring and to physical power, whether natural or artificial becomes the overwhelming desire. Why would we expect the leopard to change its spots?

You may think that the analogy with black magic or destructive magic is fanciful. If you read deep studies of magic and match it with practices in the real world then parallels are striking. Magic may intersect with religious institutions to supersede through a priesthood and assume, assert and probably ultimately invert the original mystical power that created it. Movement to a professional class with often secret ambitions for power may dislocate elements that promoted original community benefit. Power becomes contained or exercised through rituals and fixed scripts like codes. The persistent connection between cryptography and magic is evident. Magic is often about action in the world through instruments of a group in accordance with secret rituals and codes that purport to exert power. The real force may derive from the circle of power formed, the bonds between and ceremonies performed. Gangs and initiation have such rules and derive rational, collective power from the suspension of normal rules for extraordinary ambition and confinement of actions to people who can be trusted to share a common interest in pursuit. Such destructive magic is often distinct from ideas of natural magic, traditional ideas of 'witches,' the mystical approach to the imaginal world and the idea of magic as an empowering tool particularly for the individual. That does not mean individual transhumanists are practising, black or destructive magicians. It will be no surprise if they begin to suggest that they are the new witches being persecuted such is the practice of inversion. What is argued here is that materialism may become similar to notions of black magic and forces like transhumanism may embody that will, consciously or not. If you read what some transhumanists say you will see it

Some transhumanists may say that humans are zombies if they do not want to alter their body, minds and souls to the will of a machine-loving group of people who dislike other people. They may distance themselves from those posthumanists who appear to criticise the Enlightenment because transhumanists want more of the same. While denying essential qualities transhumanists believe that the one defining aspect of humanity is its quest for method and control. This is the central goal of the purification and concentration of a mentality that drives science towards scientism and magic towards destructive magic. Driven by a narrow sense of Darwinism and building on a Faustian or Promethean disposition this movement bets all on this function. Therefore it believes that technology must accentuate this quality of ratiocination only so that ultimately we can be translated into the form of the machine which we create. This is the great sacrifice that is a feature of many magical traditions and humans are the victims. If you think this is an exaggeration on my part, read what the range of transhumanists write. While some seem very sensible and some are engaging in legitimate studies of life prolongation, the perspective which drives this movement is the same that has reduced nature and humanity hitherto. It will not protect the environment. It will not protect human rights. It will not protect human liberty save for the few. It will scavenge for scraps from Christian theology and distort ideas of theosis to argue that this is somehow a spiritual and even personal process and we are spiritual machines. In doing so, it will deny or distort any traditional idea of spirituality or offer instead a pervasive, pseudo-spiritual replacement. Unfortunately transhumanism and posthumanism are part of the problem.

The nature of persons has been debated back to Aristotle and through Aquinas (1225-1274) and Boethius (480-524). However it is not confined to specific ideas of theology and represents lived experience and assumption for people in a vast range of contexts. The human person has been the target of assault by all tyrants and tyrannous systems. The people and processes that engage in such destruction do so from desires for control of others and hatred as appropriately conditioned by de-humanising forces. Whatever solution we seek will not be found in further de-humanisation. Rather we should emphasise the human person and their full efflorescence.

Consciousness is. You are. You are it. You have it. It has you. Any merely external correlation of activity and predictability in any mental model, no matter how comprehensive, is merely a reflection. If the constructor of the model of consciousness excised certain elements because they are unable to count or even perceive them then the image will be distorted. Any representation is not the thing itself. Qualia may represent the best term to describe an ineffable characteristic of experience. However, qualia are still only an attempt to indicate the seer or perceiver behind experience and capability of having experiences. We are not collections of attributes supposedly haphazardly contained by fluke in a universe without purpose oddly only comprehensible to those who accept that reductionist analysis. We must stop believing that models we create to satisfy our desperate need for comprehension, predictability and meaning can adequately address the ultimate Mystery. We will not solve the ultimate Mystery by merely denying it. Having put the problem simply let me put the answer simply.

There is a lot of discourse about the idea of the monster. Paradoxically many who see them elsewhere seem happy enough to support the idea that the human is the monster instead. Overcoming the monster is a way of telling a story, an archetypal approach that is well recognised by people who study them. Because monsters may be made up, does not mean that real monsters cannot be made up at our expense. You will soon see things that look like monsters. You will see networks almost impossible to escape from. You will see robots boss you and your life more. You will see monstrous clusters for the quiescent class that is favoured. Some called previous disintegrating clusters 'coketowns' and 'motowns.' Your town may be the new version. You will see more disorder for the mass that do not fit into the plans. You will see more antipathy to humans from humankind and machines. There is a strange paradox that those who profess a certain dislike or hatred of humanity seem to display in their attitude the de-humanising they criticise in others. Once I interviewed Professor Anthony Clare who did many interviews with accomplished people. However, he did not interview many scientists. He mentioned one aspect of an interview with the much respected forensic pathologist Bernard Knight in *In the Psychiatrist's Chair II* (1995) pg.171.

> *"Professor Knight makes no bones about the fact that he holds humanity in general in considerable contempt. As he puts it with uncharacteristic passion and venom- 'humanity stinks."*

Clare wondered whether Knight was drawn to his difficult work because of this attitude. He also pointed out that Knight loved animals 'like many passionate haters of humanity.' Knight saw humans as a 'malignancy on the face of the earth.' I too think we need to respect animals and the biosphere and cannot comprehend what such work has demanded from this man. That love of animals should not however be an elevation above humanity. In some senses the preference for animals is based on pets which do not have an independent life consistent with their nature. As we are transformed into 'pets,' in the zoo, borrowing a bondage term also, perhaps our scientific masters may 'like' us a little more, especially as we will provide great sources for the experimental curiosity and resources for the elect prolongation of life. But it seems that many would like more our disappearance justified on some sort of inherent belief that they are removed from humanity because of their calculating or objectifying disposition. If we respect people based on a deep appreciation of consciousness our disposition should manifest in greater humanity. The irony is that the hatred of humanity often comes from those who proclaim their superior vaguely scientific remove. Thus I suggest:

The antidote to total-totalitarianism of technology is suggested by subjectiveness of the person which is the object of control systems. Personalism indicates the desirability of self-actualisation in a holistic and spiritual way using phenomenology of individual investigation and qualia to have experience of highest consciousness or animasphere and through that sovereignty build solidarity through relationships that inform the communal biosphere to oppose a monstrous technosphere.

Resist Dispiriting

If all you want is escape from darkness
To leave any inner being and myth alone
To relinquish imago or conscience zone
There are many who'll help forgo the blessed.
Forget the forge within where red sparks fly
Where fact's anvil waits for smith shape
In hammered orange tumbling heartscape.
There are many who will help that tradition deny.
Amber phantom bliss of spirit would by many be
Treated like a guilty grumbling ghost
Arcane unwelcome guest to a kindly host
Archaic superstition plaintive to see.
Because to make the slave, to mint the chains
Spirit innermost cannot be allowed
Unless bellowed from below like hellfire
Save gained with searchlights to inspire a crowd
To find, authorise, vow or provoke ire
To sizzle cool and fix the delusion
Mind-cauterise with cave-shadow illusion.

About the Author

James Tunney obtained an honours degree in law from Trinity College Dublin, qualified as a Barrister at the Honorable Society of the King's Inn, Dublin and obtained an LLM from Queen Mary College, University of London.

Since then he worked as a Lecturer and Senior Lecturer in UK universities. He has been a Visiting Professor in Germany and France, lecturer around the world and worked as an international legal consultant in places such as Lesotho and Moldova for bodies such as the UNDP. He talked in many countries and published regularly on issues associated with globalisation. He has taught, written and talked about subjects such as indigenous rights, travel and tourism law, culture and heritage, intellectual property, communications technology law, competition law, China and World Trade.

He decided to leave the academic world behind to concentrate on artistic and spiritual development. He has exhibited paintings in a number of countries and has continued his writing.

You can help his work by leaving a review on Amazon.

www.ingramcontent.com/pod-product-compliance
Lightning Source LLC
Chambersburg PA
CBHW020631220526
45464CB00001B/103